Night Sky

A FIELD GUIDE TO THE CONSTELLATIONS

by Jonathan Poppele

Adventure Publications, Inc.
Cambridge, Minnesota

ACKNOWLEDGMENTS

To my brother Eric for a lifetime of inspiration.

I thank Larry Rudnick for sparking my interest in astronomy and teaching me how to teach. I am also grateful to Jon Young for his inspiration and guidance in sharing the natural world with others. A special thanks to Torsten Bronger, for his excellent PP3 software that I used to create the graphics for this book. Thanks also to Russ Durkee for reviewing this book.

Edited by Brett Ortler
Cover and book design by Jonathan Norberg

Photo Credits:
Dan Downing: 25 **Florian J. Kraemer:** 19 **NASA:** 32, 292 (bottom), 295, 296, 298 (bottom), 300 **NASA/ESA/Johns Hopkins University:** 291 **NASA/Johns Hopkins University Applied Physics Laboratory/Carnegie Institution of Washington:** 294 **NASA/JPL:** 297, 299 **NASA/JPL-Caltech/University of Wisconsin:** 290 **NASA/JPL/USGS:** 298 (bottom) **NASA/NSSDC:** 292 (top), 301 **Jonathan Poppele:** 27, 34, 40, 44, 48, 52, 56, 60, 64, 68, 72, 76, 80, 84, 88, 92, 94, 98, 102, 106, 110, 114, 118, 122, 126, 130, 134, 138, 142, 146, 150, 154, 158, 162, 164, 168, 172, 176, 180, 184, 188, 192, 196, 200, 204, 208, 212, 216, 220, 222, 226, 230, 234, 238, 242, 246, 250, 254, 258, 262, 266, 270, 274, 278, 282, 286

10 9 8 7 6 5 4 3 2 1

Copyright 2009 by Jonathan Poppele
Published by Adventure Publications, Inc.
820 Cleveland Street South
Cambridge, MN 55008
1-800-678-7006
www.adventurepublications.net
ISBN-13: 978-1-59193-229-1
ISBN-10: 1-59193-229-7

TABLE OF CONTENTS

ABOUT THIS FIELD GUIDE

Few things are as awe-inspiring as the night sky on a clear night. When we see hundreds of stars, we wonder about their origins, we marvel at the sheer number, and we can't help but be amazed that we're looking at the same stars that are visible from lands and cultures all over the world. When we think about stars, we also think about how ancient cultures observed the stars and used them to navigate, forecast the seasons and used them as a basis to theorize about their place in the world. Stargazing is a unifying, humbling experience that puts our tiny place in the universe in perspective.

Stargazing Is Accessible for Everyone

Stargazing is probably the most accessible natural experience we have. All one has to do is step outside on a clear night—no equipment is needed, and you don't have to travel. What's more, while it's difficult to find family activities that engage the young and old alike, everyone loves stargazing; three-year-olds are enamored by the night sky, and the elderly are as awe-inspired as ever.

A Focus on Constellations

When any group gathers to stargaze, discussion invariably turns to the constellations. We try to locate them, trace them with our fingers and talk about their accompanying mythology, but this often results in frustration. When you're dealing with small points of light, it's difficult to communicate what you're seeing to others and questions arise about which constellation is which. It's challenging—there are so many stars!

Night Sky **Makes Identification Easier Because**

1) each constellation is featured individually;

2) *Night Sky's* maps and graphs include well-known reference points like the Big Dipper, the North Star and the horizon;

3) constellations are organized first by season and then by how easy they are to locate.

Useful for the Casual Observer and the Expert Alike

The book is designed for everyone—the organization and design are simple and intuitive. Nonetheless, *Night Sky* includes a wealth of detailed information for the more experienced stargazer, including a reference section in the back that discusses notable deep sky objects, a look at the sun and the solar system and more.

Companion Deck of Playing Cards

We've created a companion deck of playing cards (sold separately) to help you learn the constellations. Each card features a different constellation, and the suit indicates the season in which the constellation is most prominent.

♥ = spring ♠ = summer ♣ = fall ♦ = winter

The cards are also ranked by how easy it is to spot a given constellation. Playing with the cards will help you learn the patterns of the constellations without even trying. Not only will you be having fun, but you'll find it easier to identify the constellations when you go out stargazing.

HOW TO USE THIS BOOK

Because the Earth spins on its axis and revolves around the sun, the constellations appear to move, making it sometimes difficult for novice stargazers to find them. However, by providing a few simple steps, *Night Sky* makes locating them as easy as possible.

Step 1: Choose Your Season

Each constellation is categorized into the season during which it is highest in the evening sky and, thus, easiest to see. For instance, Ursa Major (page 37) is a spring constellation because, while it is visible throughout the year, it is highest in the late evening sky in April.

Determine which season is applicable for your stargazing adventure, and turn to that section. For the purposes of this book, use the following seasons:

Spring: Mar 21–Jun 20   **Summer:** Jun 21–Sep 20
Fall: Sep 21–Dec 20   **Winter:** Dec 21–Mar 20

Step 2: Choose Your Constellation

Within each season, the constellations are organized from easiest to most difficult to locate. We recommend starting at the beginning of each section and working your way toward the end.

HINT: If you can already locate some constellations such as Ursa Major (see page 37), Ursa Minor (see page 41) or Orion (see page 223), it might be a good idea to practice with those, just to give you a better idea of how the book works.

Step 3: Choose Your Location Method

A *Where to Look: Overhead Map* and a *Where to Look: Horizon Graph* are included with each constellation. Try them both, and use the one that you prefer.

OVERHEAD MAP *(for people who prefer a "map" approach)*

a. First, face south. Then hold the book over your head.

b. The red dot at the center of each map correlates to directly overhead. The constellation is depicted by a yellow star-like symbol in relation to the Big Dipper and the North Star, which is part of the Little Dipper *(see page 41)*.

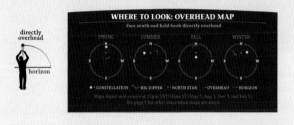

directly overhead

horizon

WHERE TO LOOK: OVERHEAD MAP
Face south and hold book directly overhead

| SPRING | SUMMER | FALL | WINTER |

● CONSTELLATION ⁓ BIG DIPPER ✶ NORTH STAR ● OVERHEAD ─ HORIZON

Maps depict mid-season at 11pm DST/10pm ST (May 5, Aug 5, Nov 5, and Feb 5).
See page 9 for other dates when maps are exact.

HORIZON GRAPH *(for people who prefer a "graph" approach)*

a. Find the current month at the bottom of the graph.

b. Look at the top of the graph above that month and face that direction. In the example below, if it were January, you would face northwest.

c. Use the star symbols to determine how high in the sky to look. Keep in mind that the bottom line is the horizon and the top line is directly overhead. In the example, in January you would look at about a 45° angle, about midway between directly overhead and the horizon.

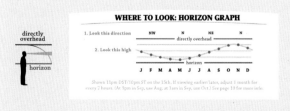

directly overhead

horizon

WHERE TO LOOK: HORIZON GRAPH

1. Look this direction   NW   N   NE   N
   directly overhead

2. Look this high

   horizon
   J F M A M J J A S O N D

Shown 11pm DST/10pm ST on the 15th. If viewing earlier/later, adjust 1 month for
every 2 hours. (At 9pm in Sep, use Aug, at 1am in Sep, use Oct.) See page 10 for more info.

7

Step 4: Determine Your Viewing Time

You can use the *Where to Look: Overhead Map* and the *Where to Look: Horizon Graph* as is at the times indicated or make modifications as follows:

ADJUSTMENTS FOR THE OVERHEAD MAP

The maps depicted will exactly match what you'll see at 11pm DST/ 10pm ST in the sky on these mid-season dates:

Spring: May 5 (midway between Mar 21–Jun 20)
Summer: Aug 5 (midway between Jun 21–Sep 20)
Fall: Nov 5 (midway between Sep 21–Dec 20)
Winter: Feb 5 (midway between Dec 21–Mar 20)

If you want to determine when the sky will match the overhead maps at other times, either:

a. Adjust for your viewing time

On the chart at the right, select your month. Use the overhead map indicated for the season that most closely references the time you are viewing.

b. Approximate the positions of the objects on the map based on movement

Study the maps on each constellation page and note how they move and rotate between each season. The rotation/movement during the seasons approximates that of the movement during the night. So if you're viewing earlier than 11pm, you'll imagine them being earlier in that cycle of movement; if you're viewing later, you'll project the movement accordingly.

In this month . . .	. . . use these maps at the times indicated			
	SPRING MAP	SUMMER MAP	FALL MAP	WINTER MAP
Early January	5am–Dawn		Dusk–7pm	11pm–1am
Late January	4–6am			10pm–12am
Early February	3–5am			9–11pm
Late February	2–4am			8–11pm
Early March	1–3am			7–9pm
Late March	1–3am*			Dusk–9pm*
Early April	12–2am*			
Late April	11pm–1am*	5am–Dawn*		
Early May	10pm–12am*	4am–Dawn*		
Late May	Dusk–11pm*	3am–Dawn*		
Early June	Dusk–10pm*	2–4am*		
Late June		1–3am*		
Early July		12–2am*		
Late July		11pm–1am*		
Early August		10pm–12am*	4am–Dawn*	
Late August		9–11pm*	3–5am*	
Early September		8–10pm*	2–4am*	
Late September		Dusk–9pm*	1–3am*	
Early October		Dusk–8pm*	12–2am*	6am–Dawn*
Late October			11pm–1am*	5am–Dawn*
Early November			9–11pm	3–5am
Late November			8–10pm	2–4am
Early December			7–9pm	1–3am
Late December	6am–Dawn		6–8pm	12–1am

* Daylight Saving Time

The seasonal sky maps on pages 34, 92, 162 and 220 are larger versions of the overhead maps. The times listed above also apply to the seasonal sky maps and indicate when to use them.

See next page for the *Where to Look: Horizon Graph* viewing time adjustments

ADJUSTMENTS FOR THE HORIZON GRAPH

The horizon graph shows the direction and height of each constellation above the horizon at 11pm DST/10pm ST on any night of the year.

It's easy to adjust the horizon graph for viewing times other than the 11pm time indicated. The general rule is simple: For every two hours earlier than 11pm DST/10pm ST, refer to the location indicated for the **previous** month. For every two hours later than 11pm DST/10pm ST, refer to the location indicated for the **next** month.

For example: If it's 9pm in September, refer to the location for August. If it's 1am in September, refer to the location for October.

Detailed Information in the Back of the Book

The content of *Night Sky* is geared toward the casual star-gazer. However, readers looking for extra detail can turn to the back of the book for a more in-depth look at stars, the solar system, tables to locate the planets and more.

Flashlight

In order to help you read this book while stargazing, we're including a red-LED flashlight. Using a red light means your eyes can adjust to night vision and stay that way while reading. Using a white light would cause you to lose your night vision. Given that full acclimation to night vision can take nearly half an hour, this should prove to be a useful tool for viewing the night sky and reading.

The Battery in This Flashlight Is Recyclable!

The flashlight included with this book contains a lithium battery (CR1220), and as with all "button" style batteries, recycling is encouraged. In some locations, disposal in household garbage is not allowed. Look for a recycling center that accepts lithium batteries. Retail stores may accept used batteries for recycling, as well as municipal or county household hazardous waste facilities.

Where This Book Works Best

Night Sky is appropriate for almost the entire continental United States and much of southern Canada. Specifically, the overhead maps and horizon graphs were created for the area that spans 50 degrees north to

Appropriate for most of the U.S. and southern Canada

30 degrees north latitude; this range includes everything from southern Canadian cities like Victoria, Ottawa, Montreal and Quebec all the way down to central Texas (Austin, Houston) and northern Florida (Tallahassee). As you move north in this range, southern constellations get lower in the sky and eventually disappear. At the same time, even more northern constellations become circumpolar (they appear to circle the pole, never dipping below the horizon) and are visible for more of the night. As you move south in this range, you will start to see stars and constellations that are not in this book.

Note: You may see the constellation in a different orientation than shown (i.e. upside down).

Star Name or
Object of Interest

IF YOU PREFER THE MAP METHOD

a. First, face south. Then hold the book over your head.

b. The red dot at the center of each map correlates to directly overhead. The constellation is depicted by a yellow star-like symbol in relation to the Big Dipper and the North Star, objects most people know.

✳ = CONSTELLATION　⤸ = BIG DIPPER
● = NORTH STAR　　＝ OVERHEAD　⌒ = HORIZON

WHERE TO LOOK: OVERHEAD MAP

Face south and hold book directly overhead

SPRING　　SUMMER　　FALL　　WINTER

✳ = CONSTELLATION　⤸ = BIG DIPPER　● = NORTH STAR　● = OVERHEAD　⌒ = HORIZON

Maps depict mid-season at 11pm DST/10pm ST (May 5, Aug 5, Nov 5 and Feb 5).
See page 9 for other dates when maps are exact.

CONSTELLATION NAME *(Pronunciation)*

English Name: English meaning or translation of the constellation name

Size: How large the constellation is, including its overall size ranking out of the 88 official constellations

sign of the **zodiac**

When to Look: The best months to see the constellation

Notes: Introductory information about the constellation, including historical background, how prominent it is in the night sky, the brightness of its stars, and how easy or difficult it is to locate and trace (make out the general outline of the constellation).

IF YOU PREFER THE GRAPH METHOD

a. Find the current month at the bottom of the graph.

b. Look at the top of the graph above that month and face that direction. In the example below, if it were January, you would face northwest.

c. Use the star symbols to determine how high in the sky to look. Keep in mind that the bottom line is the horizon and the top line is directly overhead. In the example, in January you would look at about a 45° angle, about midway between directly overhead and the horizon.

WHERE TO LOOK: HORIZON GRAPH

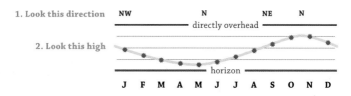

Shown 11pm DST/10pm ST on the 15th. If viewing earlier/later, adjust 1 month for every 2 hours. (At 9pm in Sep, use Aug; at 1am in Sep, use Oct.) See page 10 for more info.

Stars in the Constellation

	LIGHT YEARS	MAGNITUDE	NAME ORIGIN
Star Name (α)	72	2.1	Meaning of the star's name
Star Name (β)	88	2.2	Meaning of the star's name
Star Name (γ)	120	2.3	Meaning of the star's name

Light Years: How far the star is from Earth in light years. A light year is the distance light travels in one year in a vacuum, which is about six trillion miles.

Magnitude: How bright the star is. Every star is ranked in terms of brightness, or magnitude. The brighter the star, the lower the magnitude. Stars with negative magnitudes are very bright. The sixth magnitude is the limit of human vision; the brightest stars are first magnitude or less.

Star Name: Stars are listed by their common names. If no common name exists, its Bayer designation or its Flamsteed number is given. Bayer designations indicate roughly how bright a star is in comparison to stars in the same constellation. A Bayer designation of alpha (the first Greek letter) is given to a constellation's brightest star. Beta (the second Greek letter) is given to the constellation's second-brightest star, etc. Flamsteed numbers label stars according to their position within a constellation from west to east. The westernmost star is labeled "1," the second is labeled "2" and so on.

Objects of Interest: Additional information about the constellation's stars, including unique characteristics, historical information and fun facts. This section also includes information about other celestial objects visible in the vicinity of the constellation, including other galaxies, nebulae, star clusters and more.

Mythology/History:

Mythological and historical information about the constellation, including a detailed summary of the Greco-Roman myths commonly associated with it. For those constellations with little or no mythology associated with them, this section includes information about the constellation's discovery and discoverer and its scientific importance in the context of the history of astronomy.

In addition, the Mythology/History section includes a drawing of the person/object that the constellation represents; a similar illustration next to the "Constellation Name" and "Pronunciation" headings two pages earlier, depicts the constellation's stars, as well.

A BRIEF INTRODUCTION TO STARGAZING

Throughout history, stars have served as a compass, a clock and a calendar, and people have even tried to use them to predict the future. While most of us no longer depend on stars in these ways, stargazing is still one of the most popular and accessible activities for nature lovers around the world.

What Is a Constellation?

For thousands of years, people have looked skyward and seen shapes within the stars. These shapes became the basis for constellations, which were associated with animals, mythological figures and legends. Not surprisingly, many constellations have ancient origins. To streamline the study of astronomy, the International Astronomical Union (IAU) formalized the constellations in 1930. In all, the IAU designated 88 "official" constellations.

A recognizable pattern of stars that does not make up one of the 88 official constellations is called an asterism. The Big Dipper is one of the most famous asterisms; its stars make up one part of the official IAU constellation Ursa Major (the Great Bear). Asterisms can also stretch across several constellations. The Summer Triangle is an example; it consists of the brightest stars in the constellations Lyra, Cygnus and Aquila.

Ptolemy, Ancient Greek Astronomer

Most of today's constellations stem from ancient Greece and the Middle East. This is largely thanks to Claudius Ptolemy, a Greek astronomer who lived in Alexandria, Egypt, during the second century AD. His famous book, the *Almagest* (Latin for "Great Treatise"), was a compilation of the astronomical knowledge of his time. Drawing upon the work of Aristotle and other Greeks, Ptolemy described 48 constellations and laid out a detailed geo-centric (Earth-centered) model of the universe. The *Almagest* was the authoritative work in astronomy for 1,400 years, until the six-teenth century, when Nicolas Copernicus proposed that the sun, not the Earth, was at the center of the solar system. Galileo and other astronomers built upon this theory, leading to the eventual acceptance of the theory that the Earth (and the other planets) rotated around the sun which was only one of many stars.

TYPES OF CONSTELLATIONS

Ancient Constellations

The 48 constellations that Ptolemy helped canonize are referred to as the ancient constellations. Many of the most notable stars in the sky and several well known constellations, including Ursa Major and Orion, are among them.

Modern Constellations

By the nineteenth century, European astronomers had charted the southern skies, creating new constellations as well as adding to the Northern Hemisphere's list. These most recent additions are referred to as modern constellations.

The Constellations Today

Thanks to the IAU, the world now recognizes 88 constellations. There are no gaps between them, and every star in the sky belongs to one—and only one—constellation. Of course, not every star associated with a constellation figures into its shape. Many surrounding stars are included because of their apparent proximity. This allows for a worldwide frame of reference (a roadmap, if you will) to the night sky.

The Constellations of the Zodiac

The *Almagest* included all of the constellations of the zodiac except for Libra, which was added later. These constellations traced the progress of the sun during different times of year. The sun appeared to rise in the general vicinity of a specific constellation or "house" of the zodiac. These were not astronomically accurate and were only rough approximations of the sun's position. Since the creation of the zodiac, the Earth's axis has wobbled slightly (called precession). For these reasons, the traditional astrological dates are inaccurate. The correct dates, which are astronomically precise, include dates when the sun transitions from one constellation into another. These modern dates include a gap between Scorpio and Sagittarius; this occurs because the sun appears in a non-zodiacal constellation, Ophiucus, between them.

♈ **Aries** (pg. 193)
Traditional: Mar 21–Apr 19
Current: Apr 17–May 11

♉ **Taurus** (pg. 231)
Traditional: Apr 20–May 20
Current: May 12–Jun 18

♊ **Gemini** (pg. 235)
Traditional: May 21–Jun 20
Current: Jun 19–Jul 18

♋ **Cancer** (pg. 247)
Traditional: Jun 21–Jul 22
Current: Jul 19–Aug 7

♌ **Leo** (pg. 45)
Traditional: Jul 23–Aug 22
Current: Aug 8–Sep 14

♍ **Virgo** (pg. 53)
Traditional: Aug 23–Sep 22
Current: Sep 15–Oct 28

♎ **Libra** (pg. 65)
Traditional: Sep 23–Oct 22
Current: Oct 29–Nov 20

♏ **Scorpio** (pg. 107)
Traditional: Oct 23–Nov 21
Current: Nov 21–Nov 28

♐ **Sagittarius** (pg. 111)
Traditional: Nov 22–Dec 21
Current: Dec 16–Jan 17

♑ **Capricorn** (pg. 131)
Traditional: Dec 22–Jan 19
Current: Jan 18–Feb 13

♒ **Aquarius** (pg. 197)
Traditional: Jan 20–Feb 18
Current: Feb 14–Mar 9

♓ **Pisces** (pg. 201)
Traditional: Feb 19–Mar 20
Current: Mar 10–Apr 16

Movement of the Stars Through the Night

The daily motion of the sun across the sky is a pattern we're all familiar with. This apparent movement is caused by the rotation of the Earth on its axis. As the Earth spins, the sun appears to rise in the east, move across the sky, and set in the west. The Earth's rotation causes the stars to appear to move across the sky in the same way.

In the Northern Hemisphere all stars appear to move across the sky tracing a circle around Polaris—the North Star. Stars that are very close to Polaris trace small circles around the pole. Such stars are called "circumpolar" stars and are visible throughout the night, every night of the year. Although they make a complete circle around the pole, you will never see the full circle in one night, because the sun will rise, washing out the stars.

As we look farther from Polaris, the stars trace larger and larger circles in the sky, dipping below the horizon at the bottom and swinging far overhead at the top. These stars trace a circle that goes around us too, but this circle is interrupted by the horizon.

Time lapse star trail photo (2.5 hours)

Stars far to the south trace a circle around the south pole. To us, these stars appear to rise in the southeast, trace a low arc across the southern horizon, then set again in the southwest.

As the Earth orbits around the sun, it faces different parts of the sky. Each night, stars will appear to rise and set about 4 minutes earlier than the night before. The overhead maps and horizon graphs in this book will help you to see which constellations are visible at certain times of the year, where you can find them in the sky and the paths they take across the sky.

FREQUENTLY ASKED QUESTIONS ABOUT STARGAZING

Do I need a telescope or other expensive equipment?

No. Every object discussed in this book can be seen with the naked eye under good conditions. With that said, some objects in this book can be seen more easily with a pair of binoculars. If you own a pair, bring them along.

What else should I bring with me when I go stargazing?

Since you will spend most of your time looking up, it is nice to have a blanket and a pillow. You will also want to dress warmly, as the night air can get chilly, even in the summer. Be sure to bring the red-LED flashlight included with this book. The flashlight will help you read the book while allowing you to retain your night vision. Night vision is the result of a change in sensitivity of the light-sensing cells (rods) in your eyes. White light "resets" one's night vision, whereas red light does not. If you'd like to bring additional sources of red light, inexpensive red-LED flashlights are available. If you want to make your own, you can simply place a piece of red plastic wrap over a regular flashlight lens. And of course, be sure to bring this book!

How many stars can I see with the naked eye?

It depends on where you are. Unfortunately, many urban and suburban areas suffer from light pollution, and many stars get lost in the resulting glare. This can cause a dramatic reduction in the number of stars one can see. Under perfect conditions (areas with no light pollution, dry air and at a high altitude), one can see up to sixth-magnitude stars, which means one could theoretically see about 9,000 stars. Nevertheless, since we can only see half of the celestial sphere at one time, observers under perfect conditions will see about 4,500 stars. By contrast, in a typical rural area, one can see about 1,000 stars at one time. In the suburbs one can only expect to see 250 stars, and in an urban area only the brightest stars are visible (100 stars or less).

How many stars are there in the universe?

No one knows for sure, but the number is unimaginably large. Astronomers estimate there are approximately 200 billion stars in our galaxy alone, and many current estimates indicate there are just as many galaxies. Even in our own galaxy, most of the stars are too far away for us to see from Earth—they appear to blend together in the band of the Milky Way that stretches across the sky.

OK, so what is that really bright star I see?

If you see a very bright star that you can't find easily on one of the sky charts, it is probably a planet. The brightest objects in the night sky besides the moon are the five other planets closest to the Earth: Mercury, Venus, Mars, Jupiter and Saturn. All of these planets can be quite bright; in fact, Venus and Jupiter are always brighter than the brightest stars, and Mercury, Mars and Saturn are among the brightest objects in the sky. Because these close neighbors are moving around the sun at different rates, their position in the sky is constantly changing, and they are not shown on the fixed star charts. If you're looking for a particular planet, see the planet location tables on pages 304–308.

Why do stars twinkle?

Stars are so far away that they appear to us as only pinpoints of light, even through telescopes. Small distortions and changes in the Earth's atmosphere cause these points of light to move slightly, creating the appearance of shimmering or twinkling. Planets, by contrast, appear to us as very tiny disks. Because their light is more spread out, small distortions do not cause their apparent position to shift. For this reason, planets don't twinkle as much as stars.

How far away are the stars that I see?

Astronomers measure the distances between stars in units called light years. A light year is the distance the light travels in one Earth year—approximately 6 trillion miles. Procyon, the brightest star in Canis Minor (page 243) is quite close by astronomical standards at about 11.4 light years from Earth. Betelgeuse (pronounced *BET-el-jooz*), the second-brightest star in Orion, looks nearly as bright as Procyon, but it is about 1,400 light years away!

How do stars get their names?

The ancient civilizations of Mesopotamia, Egypt and Greece all had names for the brighter stars in the sky. Because of the preservation of ancient astronomical knowledge by Arab scholars, most of the star names inherited by contemporary science have Arabic origins or are Arabic translations. Many of these names are colorful and descriptive, such as Betelgeuse, which is pronounced *BET-el-jooz* and means "shoulder of the giant";

Sadalsud, which is pronounced *SAH-dul-su-ood* and means "luckiest of the lucky"; and Deneb Kaitos, pronounced *DEN-ebb KAY-tos*, which means "tail of the sea monster."

During the sixteenth and seventeenth centuries, two more systematic approaches to naming stars were developed. In 1603, German astronomer Johann Bayer published a star catalog using a system of Greek letters followed by the name of the constellation. Bayer assigned letters of the Greek alphabet to the stars in approximate order of brightness, with the brightest star in the constellation generally labeled as the alpha (α) star, the next brightest being the beta (β) star and so on. Over 100 years later, English astronomer John Flamsteed's

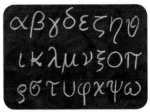

Greek alphabet in lowercase

star catalog was published, introducing another naming system. Flamsteed gave each star visible to the naked eye a number followed by the name of the constellation.

Today, stars are officially named by the International Astronomical Union, an organization of professional astronomers founded in 1919 and the international governing body for professional astronomy. For most bright stars, the IAU recognizes the ancient common names, the Bayer and Flamsteed designations and modern catalog numbers. Thus the bright star in the constellation Cygnus is called Deneb, α-Cygni (pronounced *AL-fah SIG-nye*), 50 Cygni and HD 197345. This book primarily uses the recognized common names, with reference to the Bayer or Flamsteed names as appropriate.

Can I name a star after someone?

Existing star names are fixed, and newly discovered stars are given only a catalog number. Stars cannot be named after people, and star names are not for sale. Some unscrupulous businesses "sell" star names and offer to name a star after you or a loved one. The only place such a name is ever recorded is in that business's own records and publications. These names are not official star names and are not recognized by any official organizations.

How bright is that star?

Astronomers have ranked stars by brightness since the beginning of astronomy. In fact, our current system is based roughly upon one developed by the ancient Greeks, who ranked the stars in terms of six categories, or magnitudes. Within this system, the lower a star's apparent magnitude, the brighter the star. Sixth-magnitude stars are very dim and hardly visible, whereas first-magnitude stars are very bright. With the development of photography and objective measurements of stellar magnitude, this system was formalized, as more precise measurements could be obtained. In 1856, Norman Robert Pogson defined stellar magnitude so that a first-magnitude star was defined as exactly 100 times brighter than a sixth-magnitude star. Within this standard, each magnitude is about 2.5 times brighter than the previous one. Rather than simply being classified as first or second magnitude, stars are now assigned a precise numerical value to indicate their apparent brightness. For example, the North Star (Polaris) has a magnitude of 1.97.

While the apparent magnitude system has been refined, the basic framework remains intact—*the brighter an object is, the lower its apparent magnitude.* The only difference in the modern system is that exceptionally bright objects have been given a negative apparent magnitude. Of the objects in our solar system, only the sun, the moon, Mercury, Venus, Mars, Jupiter and

Saturn reach negative magnitudes, whereas only two stars have negative magnitudes—they are Sirius (page 227) and Canopus (which isn't in the book because it's too far south to see from most of our region). All told, only about twenty stars are of the first magnitude.

How can a planet be brighter than a star?

When we refer to brightness in this book, we mean what is called "apparent magnitude." How bright an object appears to us depends on how much light it gives off and its distance from the Earth. A small, dim object that is close to Earth can appear brighter to us than a large, bright object that is much farther away. Planets are an example of this. Oftentimes, the brightest "star" in the night sky is actually a planet. Venus ranges in brightness from magnitude −3.8 to −4.6. In comparison, Sirius in the constellation Canis Major (page 227) is the brightest star in the sky, with a magnitude of −1.47. This means that Venus appears to shine between 8 and 17 times more brightly. Then again, Venus ranges from 24–160 million miles from Earth. Sirius, on the other hand, is

Venus and the moon

8.6 light years away (about 50 trillion miles). The measure of intrinsic brightness is called "absolute magnitude." It's calculated by comparing apparent magnitude to distance from Earth. Sirius has an absolute magnitude of 1.42, while Venus's is only 30 (brighter objects are given lower numbers), so Sirius actually gives off 24 billion times as much light as Venus!

What's the best way to learn the stars and constellations?

There are four effective ways to get to know the night sky.

1) Get to know the brightest stars in the sky

There are 15 first-magnitude stars visible from most of the United States. They are the first stars visible after sunset. Learning them is a great start toward finding the constellations to which they belong, as well as finding other nearby constellations. A list of the brightest stars appears on pg. 309.

2) Get to know one season of constellations

Start with one season and learn the constellations that are most prominent during that time of year. As you get to know them, it will become easier to learn new constellations as the year progresses. For your convenience, *Night Sky* and the companion deck of cards are organized by season.

3) Get to know one constellation

Familiarizing yourself with any one constellation will help you learn a lot about constellations in general. As you study the movement of that constellation throughout the year, you will learn more about *all* of the constellations' movements. You may also discover nearby constellations in the process.

4) Get to know the patterns in the stars

Simply flipping through the book or playing with the *Night Sky* companion deck of cards (sold separately) will make the patterns in the stars more familiar. As you see the patterns over and over again, you will find yourself beginning to recognize them in the night sky. Games that involve layouts of cards (such as solitaire) are the most conducive to learning the star patterns.

OTHER OBJECTS YOU MIGHT SEE

The Moon

The moon is the Earth's close celestial companion and the most prominent object in our sky after the sun. From Earth, it appears to be the same size as the sun, covering an area 0.5° across, or about the size of your pinky nail at arm's length. One of the only celestial objects ever visible during the daytime, the moon is instantly recognizable.

Viewing tips: Easy to see and unmistakable in the sky, you'll probably never confuse the moon for another object. However, the moon is so bright, especially when it is full, that it easily washes out dimmer stars. If you are out looking for dim stars and constellations, it is best to be out when the moon is not.

The Planets

All of the planets in our solar system orbit the sun in approximately the same plane. The line of this plane in the sky is called the ecliptic—the path that the sun appears to follow across the sky. Planets will always be seen very close to the ecliptic. The planet-locating tables (page 304) give the locations of the five

The moon, Jupiter and Venus

planets that can be easily seen with the naked eye through the year 2019. To locate or identify a planet, simply look up its location on the tables and follow the simple instructions there.

Shooting Stars

Shooting stars are not stars at all, but meteors—small pieces of rock, ice, dust and other debris falling to Earth from space. Meteors are traveling at tremendous speeds when they enter Earth's atmosphere. The friction generated as they move through the atmosphere causes them to burn up, leaving a trail of light. Meteors may be seen at any time, but there are a few times each year when the Earth passes through a high concentration of debris, and falling meteors are visible every few minutes. These events are called meteor showers. Few meteors reach the surface of the Earth; on the contrary, almost all burn up in the atmosphere. Those that reach the surface are referred to as meteorites.

Viewing Tips: Meteor showers happen every year, when the Earth passes through debris left behind when comets intersect Earth's orbit. Some years feature better meteor showers than others, so check the Helpful Resources section at the back of this book for more information.

Comets

Comets are small chunks of ice and dust left over from the formation of the solar system. Most comets orbit far away in a region called the Oort Cloud. Occasionally, one of these "dirty snowballs" will sweep in toward the sun on a long, elliptical orbit and can be seen from Earth. When a comet enters the inner solar system, the icy core, called the nucleus, warms up and begins to give off gas and dust, forming a large cloud around the comet called a coma, which is illuminated by the sun. Sometimes, the gas and dust will stretch out behind the comet, forming a long tail. A comet's nucleus is usually less than a mile across, but its coma may be up to 60,000 miles across and its tail may stretch out as much as 60 million miles!

Comet Hale-Bopp

It is rare that comets ever get bright enough to be seen with the naked eye, but every couple of years one appears that can be easily spotted in binoculars. Most comets that are seen from Earth disappear back into the outer solar system, not to be seen again for thousands of years. A few dozen, such as the famous Halley's Comet, have smaller orbits and return every few decades to the delight of stargazers.

Viewing tips: Check the magazines or websites referenced in the back of the book for information about when comets will be visible. Despite their dramatic-looking tails that suggest movement, comets appear still in the sky.

Deep Sky Objects

Deep sky objects are celestial objects other than stars that exist beyond our solar system, including gas clouds, clusters of stars, and other galaxies. Deep sky objects generally appear in the sky as soft, fuzzy patches and can resemble faint comets. In fact, the first catalog of deep sky objects, the now famous Messier Catalog, was originally created as a list of objects to ignore when hunting for comets in the sky.

Viewing tips: Most deep sky objects are very faint and require clear, dark skies. The individual constellation graphics show a few deep sky objects that are visible to the naked eye under good conditions.

Nebulae

Nebula is the Latin word for "cloud." Nebulae (the plural of nebula) are giant clouds of gas and dust in space. From Earth, they look like soft, fuzzy patches in the night sky. Some of these giant clouds are the birthplace of stars, and several are visible to the naked eye. The Orion Nebula is one example.

Great Orion Nebula

Star Clusters

A star cluster is a large group of stars that are bound together by gravity and move together through space. There are two kinds of star clusters: open clusters and globular clusters.

Open clusters are groups of relatively young stars that share a common origin and are still moving together through the galaxy. Several open clusters are easily visible to the naked eye. Some, like the Hyades and the Pleiades in the constellation Taurus (page 231), appear as a group of distinct stars. Others, like the Beehive Star Cluster in the constellation Cancer (page 247), blend together into a soft, fuzzy "nebulous" spot.

Globular clusters are large, ancient clusters of stars that orbit our galaxy's core. Globular clusters often include hundreds of thousands of stars. Because globular clusters are so far away, they appear very small and faint. Although there are a few that can be seen with the naked eye, it can be very difficult to distinguish them from faint stars without optical aids.

The Milky Way

Our own galaxy, the Milky Way, is made up of an estimated 200 billion stars. Most of these stars are much too dim and far away to be seen with the naked eye. But collectively, they give off a lot of light. The light of these billions and billions of stars merges together in a band of light that stretches across our sky. Easily visible to the naked eye under good, dark skies, the band of the Milky Way is shown on indi-

The Milky Way

vidual constellation graphics throughout this book. The rich star fields of the Milky Way are great places to scan with binoculars. The longer you look, the more you see.

Andromeda Galaxy

Other Galaxies

The Milky Way is made up of hundreds of billions of stars, and the universe is made up of hundreds of billions of galaxies. Like the stars of our own galaxy, most of these are much too far away for us to see, but a few can be seen with the naked eye, appearing like faint nebulae. From North America, the Andromeda Galaxy, in the constellation Andromeda (page 173), is easily seen under dark skies. At 2.5 million light years from Earth, it is the most distant object visible to the naked eye.

Satellites and Spacecraft

Over the last fifty years, thousands of satellites and spacecraft have been put into orbit. Although they are small, they are very close and can sometimes shine quite brightly when the sunlight glints off them. In fact, when the large solar panels on a large object like the International Space Station catch the light at the right angle—a phenomenon known as "satellite flare"—

it can outshine the brightest stars, cast shadows on a moonless night and can even be seen during the day. Many other satellites can be seen as "stars" that move across the sky.

International Space Station

Viewing tips: The best time to see satellites is just after dark or before dawn, when the region above the atmosphere is still illuminated by the sun. If you see a "star" drifting across the sky, it is likely a satellite. Usually, if you watch for a while, it will fade and disappear as it passes into the Earth's shadow. Satellite flares are not rare, but don't last long either. Check the websites in the back of the book for up-to-date satellite predictions.

Aircraft

High flying aircraft can sometimes be mistaken for celestial objects when they are first spotted. If you watch these long enough, however, it will soon be evident what you are seeing. Their movement across the sky is too slow to be a meteor, and too fast to be beyond our atmosphere. Unlike satellites, their light remains steady and does not brighten then dim. Also, aircraft generally have blinking lights, which distinguishes them from satellites and celestial objects.

The Aurorae

The aurorae occur when electromagnetically charged particles emitted from the sun interact with the Earth's magnetic field, causing a variety of phenomenon, including everything from a faint glow in the sky to vivid sheets of iridescent colors streaming down from above. The aurorae are most prominent at the Earth's north and south magnetic poles. The aurora in the Northern Hemisphere are called the aurora borealis, or the northern lights. In the Southern Hemisphere, they are referred to as the aurora australis, or the southern lights.

Viewing tips: Because the aurorae are caused by particles emitted from the sun, they are strongest during periods of extra-high solar activity. Check astronomy websites for solar conditions. While the northern lights are easiest to see from northern latitudes, they are occasionally visible in the southern United States.

Aurora borealis, the northern lights

THE SPRING SKY
(overhead, facing south)

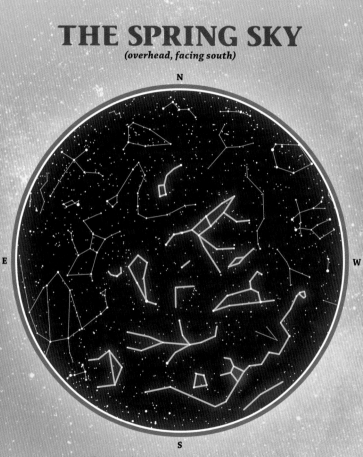

N

E   W

S

This map shows the spring sky as it appears at
1am on Apr 5, 11pm on May 5 and 9pm on Jun 5.
For other times this chart can be used, see page 9.

SPRING IS WHEN the familiar shapes of the Big Dipper (page 37) and the Little Dipper (page 41) reach their peak in the late evening sky. The first warm evenings of the year and the familiar sights in the sky invite us out to sit under the stars. The spring sky is dominated by the Big Dipper overhead, and its bowl points to the North Star and the Little Dipper.

Another prominent spring sight to see is the bright orange star Arcturus in the constellation Boötes (page 49). It is the first star visible as light fades in the evening. Other gems of the spring sky include Spica in the constellation Virgo (page 53) and Regulus in Leo (page 45). Together with the Big Dipper, these three stars frame the spring sky.

In between these prominent beacons lies a vast area of the sky with few bright stars. Look right in the center for the delicate star cluster of Coma Berenices (page 81). It is spectacular in binoculars! Many professional and amateur astronomers use this dark region of the sky as a window to the distant universe and spend spring evenings peering through telescopes at galaxies millions of light years away.

The spring constellations are in blue on the seasonal sky map to the left.

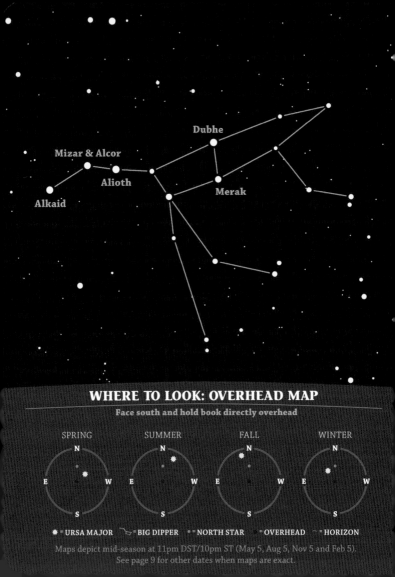

Dubhe

Mizar & Alcor

Alioth

Alkaid

Merak

WHERE TO LOOK: OVERHEAD MAP

Face south and hold book directly overhead

SPRING

SUMMER

FALL

WINTER

N
E   W
S

N
E   W
S

N
E   W
S

N
E   W
S

✳ = URSA MAJOR   ⟋ = BIG DIPPER   ◦ = NORTH STAR   ▪ = OVERHEAD   ⌒ = HORIZON

Maps depict mid-season at 11pm DST/10pm ST (May 5, Aug 5, Nov 5 and Feb 5).
See page 9 for other dates when maps are exact.

URSA MAJOR *(ER-suh MAY-jur)*

English Name: the great bear

Size: very large, 3rd largest

When to Look: most prominent from February through June and visible in the late evening sky throughout the year

Notes: Home of the best-known group of stars in the northern skies, the Big Dipper. Oldest of all the constellations, Ursa Major's history dates back at least to the last ice age. Most of the constellation, including the Dipper, is circumpolar above 40 degrees north latitude and visible every night of the year from most of the United States.

WHERE TO LOOK: HORIZON GRAPH

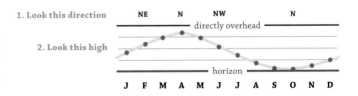

1. Look this direction

2. Look this high

Shown 11pm DST/10pm ST on the 15th. If viewing earlier/later, adjust 1 month for every 2 hours. (At 9pm in May, use Apr; at 1am in May, use Jun.) See page 10 for more info.

Stars in Ursa Major

	LIGHT YEARS	MAGNITUDE	NAME ORIGIN
Alioth (ε)	81	1.8	Arabic for "bull" or "black horse"
Dubhe (α)	124	1.8	Arabic for "bear"
Alkaid	100	1.85	Arabic for "the leader"
Mizar	78	2.2	Arabic for "girdle"
Merak (β)	79	2.3	Arabic for "loins"
Alcor	81	4.0	Arabic for "black horse"

Every star is ranked in terms of brightness, or magnitude. The brighter the star, the lower the magnitude. Stars with negative magnitudes are very bright. The sixth magnitude is the limit of human vision; the brightest stars are first magnitude or less.

Alioth (ε): The brightest star in the constellation and one of the brightest second-magnitude stars in the sky.

Dubhe & Merak (α & β): The pointer stars of the Big Dipper. A line drawn through these two stars point to Polaris, the North Star, 30 degrees distant. The name Merak derives from an Arabic word meaning "loins" (of the bear).

Alkaid: Means "the leader" in Arabic, apparently in reference to it as the leader of the three stars that form the handle of the Big Dipper.

Mizar & Alcor: This pair of stars is sometimes known as the horse and rider. Many cultures in Europe, the Middle East and North America traditionally used this pair of stars as a vision test. Mizar is a fairly bright star in the middle of the handle of the dipper. Alcor is a dimmer star just a short distance away. The pair is an optical double, lying in the same direction from Earth, but at different distances. Although Alcor is fairly easy to see with the naked eye, its close proximity to Mizar makes it easy to overlook. In fact, an older Arabic name for the star is *Suha*, meaning "the overlooked one."

Mythology/History:

The Big Dipper forms the most recognizable pattern in the northern skies. These stars have been recognized as many things: a dipper, a plow, a coffin, a wagon, a skunk, even a shark. Most commonly these stars are known as the Great Bear.

The Great Bear may be the oldest of all the constellations, dating back at least to the last ice age and possibly to a Paleolithic bear cult 50,000 years ago. Bears, however, do not have long tails, and there were different explanations for this in each myth. In many cultures, the three stars we recognize as the handle of the dipper were considered to be three bear cubs or three hunters trailing the Great Bear.

In classical mythology there is a different explanation: Ursa Major is identified with Callisto, a beautiful and renowned huntress. Zeus seduced her and fathered her child, Arcas. Hera, Zeus's jealous wife, transformed Callisto into a bear, robbing her of her beauty. When Arcas was older, he became a hunter and saw Callisto. Unaware of her true identity, he gave chase. Callisto fled into the Arcadian temple of Zeus, unaware that the law punished trespassers with death. Seeing her plight, Zeus hoisted her into heaven by her tail, stretching it. Arcas was transformed into a bear (Ursa Minor), so he could be with his mother. In other versions, Arcas is said to be the constellation Boötes.

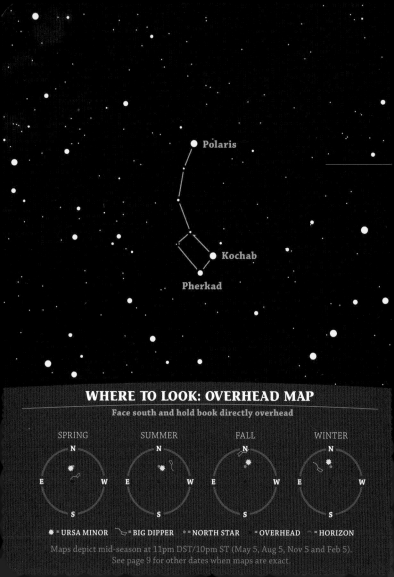

Polaris

Kochab

Pherkad

WHERE TO LOOK: OVERHEAD MAP

Face south and hold book directly overhead

SPRING   SUMMER   FALL   WINTER

✱ = URSA MINOR   〜 = BIG DIPPER   • = NORTH STAR   = OVERHEAD   = HORIZON

Maps depict mid-season at 11pm DST/10pm ST (May 5, Aug 5, Nov 5 and Feb 5).
See page 9 for other dates when maps are exact.

URSA MINOR *(ER-suh MY-ner)*

English Name: the little bear

Size: small, 56th largest

When to Look: most prominent from May through July and visible in the late evening sky throughout the year

Notes: Commonly known as the Little Dipper, the constellation includes the North Star (Polaris), the most well-known star in the sky. Polaris lies in the direction of the Earth's axis and marks the point around which all the other stars appear to rotate. Polaris never moves in the sky and is visible all night, every night of the year from everywhere in the Northern Hemisphere.

WHERE TO LOOK: HORIZON GRAPH

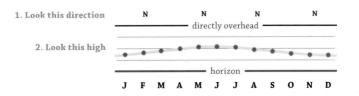

1. **Look this direction**

2. **Look this high**

Shown 11pm DST/10pm ST on the 15th. If viewing earlier/later, adjust 1 month for every 2 hours. (At 9pm in May, use Apr; at 1am in May, use Jun.) See page 10 for more info.

Stars in Ursa Minor

	LIGHT YEARS	MAGNITUDE	NAME ORIGIN
Polaris (α)	430	2.0	Latin for the "pole star"
Kochab (β)	126	2.1	Arabic name of unknown origin
Pherkad (γ)	480	3.1	Arabic for "two calves"

Every star is ranked in terms of brightness, or magnitude. The brighter the star, the lower the magnitude. Stars with negative magnitudes are very bright. The sixth magnitude is the limit of human vision; the brightest stars are first magnitude or less.

Polaris (α): The best-known star in the sky, it is also called the North Star, the Lodestar, Stella Maris, Star of the Sea, *L'Etoile du Nord*, Cynosure and many other names. It lies less than 1 degree away from the celestial pole and appears to be the one unmoving point that all the other stars rotate around. It has been recognized as the pole star since the time of the Greeks and has historically been one of the most important stars for navigation.

Kochab & **Pherkad** (β & γ): These two stars are known together as the guardians of the pole. The brighter of the two is Kochab, an Arabic name of unknown origin. It has a magnitude of 2.1, nearly equal to that of Polaris, and lies 95 light years from Earth. Beside Kochab is Pherkad; its name is Arabic for "two calves" and originally referred to both of these stars. Pherkad has a magnitude of 3.1 and is 225 light years from Earth.

The Little Dipper: The stars of the Little Dipper make a useful tool for rating your viewing conditions. The stars of the cup have visual magnitudes of approximately 2, 3, 4 and 5, allowing you to quickly determine the "limiting magnitude" for your stargazing. If you can clearly see all four stars, you should be able to see all the stars on the charts in this book.

Mythology/History:

For centuries this has been one of the most important constellations for navigation, as it includes the pole star, Polaris. Nevertheless, at the height of Egyptian culture 4,000 years ago, the star Thuban in the constellation Draco marked the north pole and the stars of Ursa Minor were of relatively little importance. At that time, these stars were considered to be part of Draco and formed an asterism known as the Dragon's Wing.

When these stars were first identified as a separate constellation by the Greeks, they were identified with the seven nymphs who tended Hera's garden where the golden apples were kept (page 121). It was the astrono-mer Thales, who lived from 625 BC to 545 BC, who first introduced the Little Bear to the Greeks. There are two primary myths associated with Ursa Minor. In one myth, the stars are said to represent Arcas, Callisto's son, who was transformed into a bear and hoisted into the heavens along with his mother (page 39).

In the other myth, the constellation is said to be the image of the nymph Cynosura, who nursed the infant Zeus. It is said that Zeus placed her in the heavens in gratitude for her care of him. The word *cynosura* is Greek for "dog's tail" and is an alternate name for both the constellation Ursa Minor and for the North Star. It is also the root of the English word "cynosure," which means "one that directs or guides."

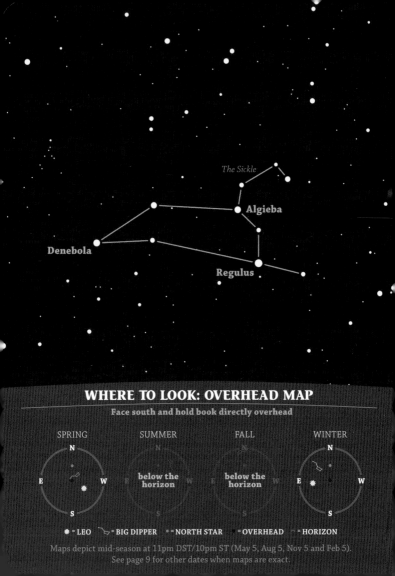

The Sickle

Algieba

Denebola

Regulus

WHERE TO LOOK: OVERHEAD MAP

Face south and hold book directly overhead

SPRING	SUMMER	FALL	WINTER
N			N
E / W	below the horizon	below the horizon	E / W
S			S

☀ = LEO ⌐ = BIG DIPPER ✷ = NORTH STAR ▪ = OVERHEAD ▪ = HORIZON

Maps depict mid-season at 11pm DST/10pm ST (May 5, Aug 5, Nov 5 and Feb 5).
See page 9 for other dates when maps are exact.

LEO *(LEE-oh)*

zodiac

English Name: the lion

Size: large, 12th largest

When to Look: most prominent from February through May and visible in the late evening sky from January through June

Notes: A prominent constellation of the zodiac. Easy to spot and fairly easy to trace, it actually looks something like a lion. The stars of the head and neck of the lion form an asterism known as "The Sickle." The constellation's brightest star, Regulus, lies just off of the ecliptic (the sun's apparent path across the sky throughout the year) and is frequently seen near the moon or one of the planets.

WHERE TO LOOK: HORIZON GRAPH

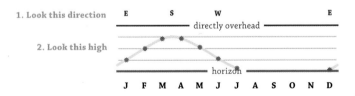

1. Look this direction   E   S   W   E

directly overhead

2. Look this high

horizon

J F M A M J J A S O N D

Shown 11pm DST/10pm ST on the 15th. If viewing earlier/later, adjust 1 month for every 2 hours. (At 9pm in May, use Apr; at 1am in May, use Jun.) See page 10 for more info.

Stars in Leo

	LIGHT YEARS	MAGNITUDE	NAME ORIGIN
Regulus (α)	85	1.4	Latin for "little king" or "prince"
Denebola (β)	39	2.1	Arabic for "lion's tail"
Algieba (γ)	126	2.2	Arabic for "brow"

Every star is ranked in terms of brightness, or magnitude. The brighter the star, the lower the magnitude. Stars with negative magnitudes are very bright. The sixth magnitude is the limit of human vision; the brightest stars are first magnitude or less.

Regulus (α): With a magnitude of 1.4, Regulus is the 15th-brightest star visible from our region and the dimmest of all the first-magnitude stars. Of the first-magnitude stars, it lies closest to the ecliptic and is often seen very near the moon and the planets.

Denebola (β): This star marks the eastern end of the constellation and one corner of the triangle that forms the lion's hindquarters. It also marks the western corner of the large asterism known as the Virgin's Diamond, together with Cor Caroli in Canes Venatici, Arcturus in Boötes and Spica in Virgo.

Algieba (γ): Arabic for "brow," this star actually marks the nape of the neck in most depictions of the constellation. It has a magnitude of 2.2 and lies 126 light years from Earth. With very sharp eyesight, or a pair of binoculars, you can separate the nearby fifth-magnitude star, 40 Leonis, which lies nearly in the same line but is only 70 light years away.

The Sickle: The stars of the lion's head and shoulders form a backward question mark or sickle that is easy to trace. Regulus marks the handle of the sickle and the dot below the backward question mark.

Ecliptic: The sun passes through Leo from August 8 to September 14. The third constellation in the zodiac, Leo's astrological period runs from July 23 to August 22.

Mythology/History:

An ancient constellation that has its origins in a constellation of a lion from the Babylonian zodiac. When the Babylonian Empire was flourishing, the sun was in Leo during the summer solstice. The association between lions and the sun remained strong in western culture for thousands of years.

The lion, often engaged in a struggle with a heroic warrior, was a common theme in Babylonian artwork. These themes were carried forward by the Greeks, who usually associated this constellation with the Nemean Lion slain by Hercules. In Greek mythology, the Nemean Lion was a monstrous creature that inhabited the countryside around Corinth. Some ancient authors say that he was the offspring of the monsters Typhon and Echidna and a sibling of many other monsters in Greek mythology. Other sources say that he fell from the moon—a story that suggests an astronomical origin for these myths.

As the first of his famous Twelve Labors (page 117), Hercules was charged with slaying the Nemean Lion. The beast's skin was said to be so strong that no blade could cut it and no arrow could pierce it. Unable to use arrows, Hercules wrestled the great lion and strangled it to death. The hero then used the lion's own claws to skin it and wore its pelt as his armor. The zodiac symbol for Leo (♌) represents the head and mane of the lion.

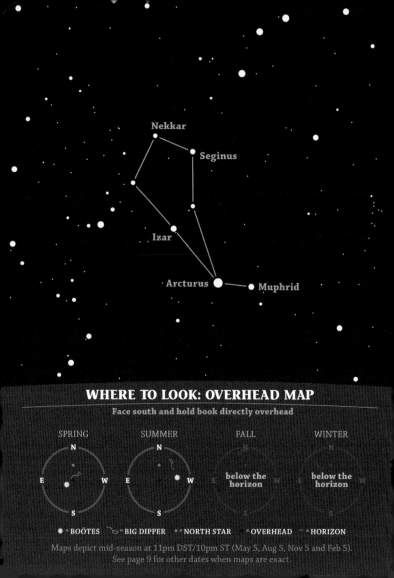

Nekkar

Seginus

Izar

Arcturus   Muphrid

WHERE TO LOOK: OVERHEAD MAP

Face south and hold book directly overhead

SPRING	SUMMER	FALL	WINTER
N	N	N	N
E · W	E · W	**below the horizon**	**below the horizon**
S	S	S	S

✹ = BOÖTES   ⌐ = BIG DIPPER   •• = NORTH STAR   •• = OVERHEAD   ⌒ = HORIZON

Maps depict mid-season at 11pm DST/10pm ST (May 5, Aug 5, Nov 5 and Feb 5).
See page 9 for other dates when maps are exact.

BOÖTES (bo-OH-teez)

English Name: the herdsman

Size: large, 13th largest

When to Look: most prominent from April through July and visible in the late evening sky from February through August

Notes: An ancient constellation highlighted by Arcturus, the brightest star in the northern skies. Easy to spot, its lopsided kite-like shape is relatively easy to trace as it stretches out above Arcturus. Boötes is easy to find when using the Big Dipper; simply follow the arc of the handle of the Big Dipper and "arc to Arcturus" and then follow that line to "speed on to Spica," a star in Virgo.

WHERE TO LOOK: HORIZON GRAPH

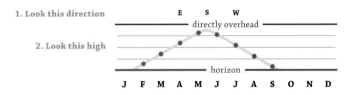

1. Look this direction

E S W

directly overhead

2. Look this high

horizon

J F M A M J J A S O N D

Shown 11pm DST/10pm ST on the 15th. If viewing earlier/later, adjust 1 month for every 2 hours. (At 9pm in May, use Apr; at 1am in May, use Jun.) See page 10 for more info.

Stars in Boötes

	LIGHT YEARS	MAGNITUDE	NAME ORIGIN
Arcturus (α)	36	−0.04	Greek for "guardian of the bear"
Izar (ε)	150	2.4	Arabic for "loincloth"
Muphrid (η)	37	2.7	Arabic for "solitary one"
Seginus (γ)	85	3.0	The result of translation errors
Nekkar (β)	220	3.5	Arabic for "ox-driver"

Every star is ranked in terms of brightness, or magnitude. The brighter the star, the lower the magnitude. Stars with negative magnitudes are very bright. The sixth magnitude is the limit of human vision; the brightest stars are first magnitude or less.

Arcturus (α): The second-brightest star in our sky and the brightest star north of the celestial equator. The name means "guardian of the bear" in Greek *arktos* being the Greek word for bear and the Greek name for the constellation Ursa Major. The association of these constellations with the north is also the root of the word "arctic." Thus, the name has also come to mean "guardian of the north." At magnitude −0.04, Arcturus is bright enough for its red-orange color to be clearly visible.

Izar (ε): The second-brightest star in the constellation, its name means "loincloth" in Arabic.

Muphrid (η): A close neighbor of Arcturus, lying just 3.2 light years away. From a planet orbiting Muphrid, Arcturus would shine in the sky with a magnitude of −5.2. That is about twice as bright as the planet Venus and bright enough to be visible during the daytime and to cast shadows on a moonless night.

Seginus (γ): This star's peculiar name is the Latinization of the Arabic form of the Greek name for the constellation.

Nekkar (β): The Arabic name for the entire constellation, it comes from the word for "ox-driver."

Mythology/History:

An ancient constellation that dates back to Babylonian times, it is associated with a wealth of Greek mythology. In one myth, Boötes is said to be the son of Demeter, the goddess of agriculture; he was placed in the sky to honor his invention of the plow. The stars of the Big Dipper are sometimes said to be the plow.

In another myth, Boötes is said to be a grape grower named Icarius who was shown the secret of wine-making by the god Dionysus. Enchanted with the beverage, Icarius loaded his ox-driven cart with wine and set out to share it with shepherds of the region. They all drank too much and awoke with hangovers. Thinking that Icarius had tried to poison them to steal their flocks, the shepherds murdered Icarius. Dionysus placed him in the stars to amend his wrongful death. In this story, Virgo is sometimes said to represent Icarius's grieving daughter and Canis Minor, his faithful dog.

In a third story, Boötes is said to be Arcas, son of Zeus and Callisto. Callisto was transformed into a bear by Zeus's jealous wife, Hera. When Arcas was hunting, he spotted the bear and gave chase, unaware it was his mother. Zeus rescued Callisto by lifting her into the stars where she became Ursa Major. Later, Arcas was also placed there to guard his mother and herd her around the celestial pole.

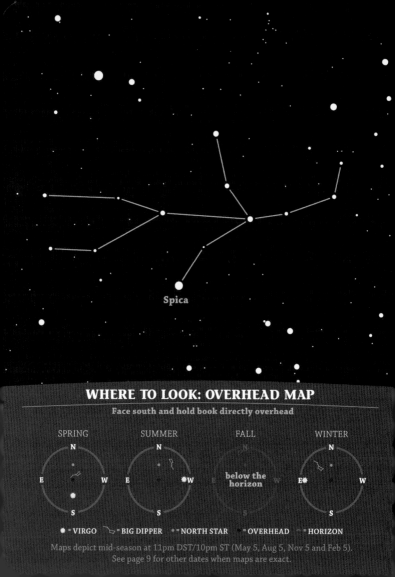

Spica

WHERE TO LOOK: OVERHEAD MAP

Face south and hold book directly overhead

SPRING	SUMMER	FALL	WINTER

below the horizon

✷ = VIRGO   ⌐ = BIG DIPPER   • = NORTH STAR   ● = OVERHEAD   ⌒ = HORIZON

Maps depict mid-season at 11pm DST/10pm ST (May 5, Aug 5, Nov 5 and Feb 5).
See page 9 for other dates when maps are exact.

VIRGO *(VER-go)*

zodiac

English Name: the virgin (maiden)

Size: very large, 2nd largest

When to Look: most prominent in May and visible in the late evening sky from March through July

Notes: The largest constellation in the zodiac and the second-largest constellation overall. Easy to find because of its bright star, Spica, which lies very close to the ecliptic and is often seen near the moon or one of the planets. The rest of this huge constellation is a bit tricky to trace as it is quite spread out and made mostly of third-and fourth-magnitude stars. Spica also marks the southern point of the large spring asterism, the "Virgin's Diamond." To find Virgo, trace the arc of the handle of the Big Dipper to the star Arcturus in Boötes ("arc to Arcturus") then follow that line to Spica ("speed on to Spica").

WHERE TO LOOK: HORIZON GRAPH

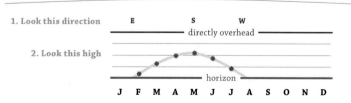

1. Look this direction

2. Look this high

Shown 11pm DST/10pm ST on the 15th. If viewing earlier/later, adjust 1 month for every 2 hours. (At 9pm in May, use Apr; at 1am in May, use Jun.) See page 10 for more info.

Stars in Virgo

	LIGHT YEARS	MAGNITUDE	NAME ORIGIN
Spica (α)	250	1.0	Latin for "ear of wheat"

Every star is ranked in terms of brightness, or magnitude. The brighter the star, the lower the magnitude. Stars with negative magnitudes are very bright. The sixth magnitude is the limit of human vision; the brightest stars are first magnitude or less.

Spica (α): One of the oldest star names still in use, it means "ear of wheat" in Latin. The 10th-brightest star in our sky, with a magnitude of 1.0, its bluish color is visible to the naked eye. It stands out sharply in the sky because there are no other bright stars nearby. Spica is the star that the Greek astronomer Hipparchus used to discover the procession of the equinoxes caused by the wobbling of the Earth's axis. In 3,200 BC, a temple was built in Thebes oriented toward Spica. A little over 3,000 years later, Hipparchus noted a significant change in the position of Spica relative to the temple. He used this, and other calculations, to estimate the procession of the equinoxes at 1 degree every 100 years—remarkably close to the actual rate of 1 degree every 72 years.

Ecliptic: Virgo is the largest zodiacal constellation, and the sun spends more time there than in any other constellation, from September 15 until October 28. The sixth constellation in the zodiac, its astrological period begins on August 23 and runs to September 22.

Mythology/History:

A constellation that dates back to the Babylonians, Virgo has been associated with nearly every prominent goddess and maiden in mythology. The stars have been said to represent the goddesses Hera, Isis, Athena and even the Virgin Mary.

In Greek mythology, she is most commonly identified with Dikē, the goddess of justice. According to legend, Dikē was the daughter of Zeus and Themis, the goddess of divine law. Dikē lived among the human race and ruled over them during the golden age of man, when people "lived like gods without sorrow of heart, remote and free from toil and grief." But after the golden age, people became greedy and morality disappeared. Civilization deteriorated into the silver age, the bronze age and finally the iron age, symbolized by the least valuable of metals. In it, justice, peace and order gave way completely to greed, war and strife. Men drew boundaries to define nations; children were born into a life of struggle and toil; and truth, modesty, and shame at wrongdoing disappeared. Upon seeing this, Dikē decided to live in heaven with the other gods.

The astrological symbol for Virgo (♍) is an ancient symbol for medicine. Its closed loop distinguishes it from the symbol for Scorpio (♏) and represents the feminine, nurturing aspect of healing.

Blaze Star (T) •-•-•-• **Gemma**

SPRING   SUMMER   FALL   WINTER

below the horizon   **below the horizon**

✴ = CORONA BOREALIS   ⌐⌐ = BIG DIPPER   • = NORTH STAR   • = OVERHEAD   ⎯ = HORIZON

Maps depict mid-season at 11pm DST/10pm ST (May 5, Aug 5, Nov 5 and Feb 5).
See page 9 for other dates when maps are exact.

CORONA BOREALIS *(cuh-ROE-nuh bor-ee-AL-iss)*

English Name: the northern crown

Size: very small, 73rd largest

When to Look: most prominent from May through July and visible in the late evening sky from March through September

Notes: A small, beautiful constellation that, with just a bit of imagination, does in fact look like a jeweled tiara. Though it contains only one bright star, its other stars are close together making it easy to trace when the skies are dark enough.

WHERE TO LOOK: HORIZON GRAPH

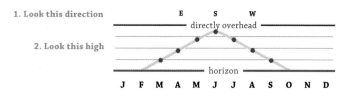

1. Look this direction

2. Look this high

Shown 11pm DST/10pm ST on the 15th. If viewing earlier/later, adjust 1 month for every 2 hours. (At 9pm in Jun, use May; at 1am in Jun, use Jul.) See page 10 for more info.

Stars in Corona Borealis

	LIGHT YEARS	MAGNITUDE	NAME ORIGIN
Gemma (α)	78	2.2	Latin for "gem"
Blaze Star (T)	6,500	10*	A modern nickname

Every star is ranked in terms of brightness, or magnitude. The brighter the star, the lower the magnitude. Stars with negative magnitudes are very bright. The sixth magnitude is the limit of human vision; the brightest stars are first magnitude or less.

* Occasionally flares to 2.0

Gemma (α): The name is Latin for "gem." The brightest star in the constellation, it is the jewel in the crown.

Blaze Star (T): Nicknamed the Blaze Star. A tenth-magnitude star, it is normally invisible to the naked eye, binoculars, or even a small telescope. This star, however, is a member of a rare class of stars called recurrent novas. Recurrent novas are star systems that are usually very faint but erupt from time to time, increasing their brightness several hundredfold for several months before dimming again. T Coronae Borealis (the formal name for the star) is one of only two recurrent novas ever visible to the naked eye (the other is RS Ophiuchi). It erupted in 1866 and again in 1946, each time reaching a magnitude of about 2.0. Another eruption could come at any time.

58

Mythology/History:

This constellation is associated with the crown worn by Ariadne, daughter of King Minos of Crete, at her wedding to Theseus. According to legend, King Minos attacked Athens to avenge the murder of his son. The Cretans defeated the Athenians and forced them to pay a bloody tribute to Minos—every nine years, seven men and seven women from Athens were sacrificed to the Minotaur, a monster with a bull's head and a man's body. It lived at the center of the Labyrinth—an inescapable maze designed to lead victims right to the Minotaur. During the third tribute, Theseus, an Athenian, volunteered to enter the Labyrinth to kill the Minotaur. Ariadne fell in love with him and conspired to aid him.

When Theseus entered the Labyrinth, Ariadne gave him his sword and a ball of string with one end tied to the maze entrance. Theseus slew the Minotaur and followed the string, leading his fellow Athenians out of the Labyrinth. Theseus and Ariadne sailed off to the island of Dia and were married.

The gods gave Ariadne a crown made by Hephaestus, the blacksmith of the gods. Made out of gold and Indian gems, it was said to glow with a light of its own. After the wedding, Theseus threw the crown into the sky, where its jewels became stars.

Chara

Cor Caroli

WHERE TO LOOK: OVERHEAD MAP

Face south and hold book directly overhead

SPRING   SUMMER   FALL   WINTER

below the horizon

✳ = CANES VENATICI   ◝ = BIG DIPPER   • = NORTH STAR   ▪ = OVERHEAD   ⌒ = HORIZON

Maps depict mid-season at 11pm DST/10pm ST (May 5, Aug 5, Nov 5 and Feb 5).
See page 9 for other dates when maps are exact.

CANES VENATICI *(CANE-eez ve-NAT-iss-eye)*

English Name: the hunting dogs

Size: medium, 38th largest

When to Look: most prominent from March through June and visible in the late evening sky from January through August

Notes: A modern constellation consisting of just two primary stars, which sit below the handle of the Big Dipper. While neither star is especially bright, the constellation is fairly easy to find. Its brighter star, the third-magnitude Cor Caroli, forms the northern point of the large spring asterism called the "Virgin's Diamond."

WHERE TO LOOK: HORIZON GRAPH

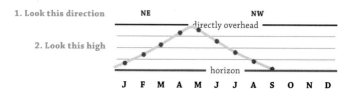

1. Look this direction

2. Look this high

Shown 11pm DST/10pm ST on the 15th. If viewing earlier/later, adjust 1 month for every 2 hours. (At 9pm in May, use Apr; at 1am in May, use Jun.) See page 10 for more info.

Stars in Canes Venatici

	LIGHT YEARS	MAGNITUDE	NAME ORIGIN
Cor Caroli (α)	110	2.9	Latin for "Heart of Charles"
Chara (β)	27	4.3	Greek for "joy"

Every star is ranked in terms of brightness, or magnitude. The brighter the star, the lower the magnitude. Stars with negative magnitudes are very bright. The sixth magnitude is the limit of human vision; the brightest stars are first magnitude or less.

Cor Caroli (α): The name, which means "Heart of Charles" in Latin, was brought into use by Edmond Halley as a tribute to King Charles I of England after the overthrow of the monarchy in 1649. The star marks the northern end of a large asterism known as the Virgin's Diamond, which also includes Arcturus in Boötes, Spica in Virgo, and Denebola in Leo.

Chara (β): The dimmer of the two "hounds," its name means "joy" in Greek.

Mythology/History:

A modern constellation created by Johannes Hevelius in the late seventeenth century, Canes Venatici is said to represent the herdsman Boötes' two hunting dogs.

The two bright stars that mark the constellation have been known since antiquity and were charted by Ptolemy among the "unformed" stars of Ursa Major. Their association with Boötes and his dogs, however, came much later. Ancient texts often described Boötes as carrying a cudgel (a short club), called a *kollorobos* in Greek. When Ptolemy's *Almagest* was translated into Arabic, this was incorrectly interpreted as the similar sounding word *kalaurops*, meaning "shepherd's staff." The Arabic translations used the phrase *dhāt al-kullāb*, meaning "shaft with a hook," to describe this. When the Arabic texts were translated into Latin, the translators mistook the Arabic word *kullāb* for the word *kilāb*, meaning dogs, and rendered it as *hastile habens canes*, or "shaft with dogs." This odd expression led several astronomers to look for Boötes' dogs.

Several incarnations of these dogs were proposed before Hevelius formalized his new constellation in his star atlas of 1690.

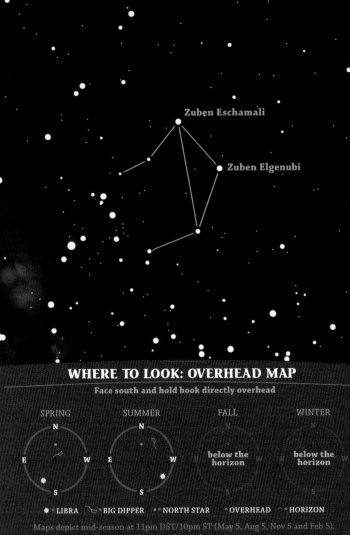

Zuben Eschamali

Zuben Elgenubi

WHERE TO LOOK: OVERHEAD MAP

Face south and hold book directly overhead

SPRING	SUMMER	FALL	WINTER
		below the horizon	below the horizon

☀ = LIBRA   ⌐ = BIG DIPPER   ∙∙ = NORTH STAR   ∘ = OVERHEAD   ⌒ = HORIZON

Maps depict mid-season at 11pm DST/10pm ST (May 5, Aug 5, Nov 5 and Feb 5).
See page 9 for other dates when maps are exact.

LIBRA *(LEE-bruh)*

zodiac

English Name: the balance

Size: medium, 29th largest

When to Look: most prominent in June and visible in the late evening sky from April through July

Notes: One of the constellations of the zodiac, Libra consists mostly of dimmer stars, and its shape is rather plain. These stars were once considered to be the claws of Scorpius and are easy to locate when the brilliant stars of the scorpion are also visible in the sky.

WHERE TO LOOK: HORIZON GRAPH

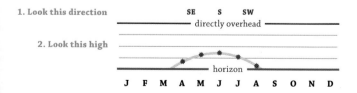

Shown 11pm DST/10pm ST on the 15th. If viewing earlier/later, adjust 1 month for every 2 hours. (At 9pm in May, use Apr; at 1am in May, use Jun.) See page 10 for more info.

65

Stars in Libra

	LIGHT YEARS	MAGNITUDE	NAME ORIGIN
Zuben Eschamali (β)	160	2.6	Arabic for "northern claw"
Zuben Elgenubi (α)	72	2.8	Arabic for "southern claw"

Every star is ranked in terms of brightness, or magnitude. The brighter the star, the lower the magnitude. Stars with negative magnitudes are very bright. The sixth magnitude is the limit of human vision; the brightest stars are first magnitude or less.

The bright stars in Libra still carry their names from when they were a part of Scorpius.

Zuben Eschamali (β): The brightest star in the constellation, with a magnitude of 2.6. Its name means "northern claw" in Arabic.

Zuben Elgenubi (α): Slightly dimmer than the constellation's beta star, with a magnitude of 2.8. Its name is Arabic for "southern claw." It lies at a distance of 72 light years.

Ecliptic: The sun passes through Libra from October 29 to November 20. The seventh constellation in the zodiac, its astrological period runs from September 23 to October 22.

Mythology/History:

Lying along the ecliptic, the stars of Libra have been recognized since Babylonian times, but the current incarnation of Libra is relatively new. In fact, among constellations of the zodiac, Libra is the mostly recently revised constellation, as this version dates only to the first century AD.

For most of their history, these stars were considered to be the claws of the scorpion in Scorpius (page 107). The Romans were the first to define Libra as a separate constellation. Around the first century AD, the sun was in Libra during the fall equinox, and day and night were in balance. The Romans recognized this by associating these stars not with the scorpion but with the great scales of justice, held by the goddess Astraeia. The stars were thereafter associated with Virgo, whom the Romans considered to be the image of Astraeia. Because of these relatively recent changes, these stars have no classical myths associated with them.

Libra is unique for being the only constellation of the zodiac not to depict a living being. The word "zodiac" comes from the Latin *zōdiacus*, meaning "circle of animals." Libra's astrological sign (♎) is a symbol of balance and is sometimes said to represent the scales of justice.

R Hydrae

Alphard

WHERE TO LOOK: OVERHEAD MAP

Face south and hold book directly overhead

SPRING	SUMMER	FALL	WINTER
	below the horizon	**below the horizon**	

☀ = HYDRA   〰 = BIG DIPPER   • • = NORTH STAR   ◆ = OVERHEAD   ⌢ = HORIZON

Maps depict mid-season at 11pm DST/10pm ST (May 5, Aug 5, Nov 5 and Feb 5).
See page 9 for other dates when maps are exact.

HYDRA *(HIGH-druh)*

English Name: the sea serpent

Size: very large, the largest constellation in the sky

When to Look: most prominent in April and visible in the late evening sky from February through May

Notes: The largest constellation in the sky and by far the longest. It stretches so far east to west that it takes six hours for the entire constellation to rise. Because Hydra is so large, it is shown here at 65% the size of all the other constellations in this book. Its enormous size makes it very difficult to see all at once, and it's best to just look for one end or the other.

WHERE TO LOOK: HORIZON GRAPH

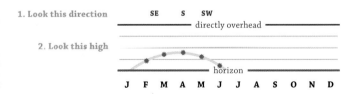

1. Look this direction
2. Look this high

Shown 11pm DST/10pm ST on the 15th. If viewing earlier/later, adjust 1 month for every 2 hours. (At 9pm in May, use Apr; at 1am in May, use Jun.) See page 10 for more info.

Stars in Hydra

	LIGHT YEARS	MAGNITUDE	NAME ORIGIN
Alphard (α)	175	2.0	Arabic for "the solitary one"
R Hydrae	2,000	7.1*	No common name

Every star is ranked in terms of brightness, or magnitude. The brighter the star, the lower the magnitude. Stars with negative magnitudes are very bright. The sixth magnitude is the limit of human vision; the brightest stars are first magnitude or less.

* Varies from 3.2 to 11.0

Alphard (α): With a magnitude of 2.0, it is the only bright star in the constellation. It is also the only star in the constellation that has a common name. Fittingly, its name means "the solitary one" in Arabic.

R Hydrae: The third of the Mira-type variable stars (page 186) to be discovered. Over a period of about 13 months, it varies from magnitude 3.2 to 11.0 and back, and it lies at a distance of about 2,000 light years from Earth. Variable stars are labeled "R" if the star is the first variable in the constellation. The next is labeled "S" and so on. This naming convention was adopted because the preceding letters in the alphabet were already used to name other types of astronomical objects.

Mythology/History:

The largest of all the constellations, Hydra was recognized by the Greeks as a water snake and is associated with two myths. In one, these stars represent the water snake in the tale of the crow and the fig tree. In this story, Apollo asked his companion bird, the crow, to fetch water from a sacred spring for a ritual. The crow brought the water, but too late, as it stopped to eat at a fig tree. He blamed his delay on a water snake that he found near the spring. Apollo saw through the lie and condemned the crow to a lifetime of eternal thirst.

Other writers associated these stars with the Lernean Hydra, a fearsome serpent with many heads and poison breath. Hercules was given the task of killing it. The hero covered his nose and mouth and used a harvester's sickle to cut off the beast's many heads. Every time Hercules cut off a head, two grew back in its place; Hercules solved this problem by using a torch to cauterize each stump, preventing it from growing back.

While Hercules battled the Hydra, the goddess Hera sent a tiny crab to distract him and tip the balance in the monster's favor, but Hercules crushed it. (This crab was immortalized as the constellation Cancer.) The Hydra's last head was immortal. Hercules cut it off and placed it under a rock. In the sky, the crab sits nearby the Hydra's head and Hercules' sickle shines just above the Hydra's neck.

Algorab

Gienah

Kraz

Minkar

WHERE TO LOOK: OVERHEAD MAP

Face south and hold book directly overhead

SPRING   SUMMER   FALL   WINTER

N   **below the horizon**   **below the horizon**   **below the horizon**

E   W

S

✳ = CORVUS   ⌐ = BIG DIPPER   • = NORTH STAR   = OVERHEAD   = HORIZON

Maps depict mid-season at 11pm DST/10pm ST (May 5, Aug 5, Nov 5 and Feb 5).
See page 9 for other dates when maps are exact.

CORVUS *(COR-vus)*

English Name: the crow

Size: very small, 70th

When to Look: most prominent in April and May and visible in the late evening sky from March through June

Notes: A small constellation distinguished by four main stars that form an irregular tetrahedron sometimes called the "sail" by mariners and the "tent" by the ancient Arabs. Though it contains no truly bright stars, it is quite compact and easy to trace.

WHERE TO LOOK: HORIZON GRAPH

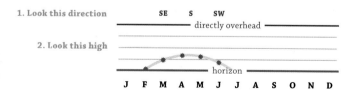

1. Look this direction

SE   S   SW
directly overhead

2. Look this high

horizon

J F M A M J J A S O N D

Shown 11pm DST/10pm ST on the 15th. If viewing earlier/later, adjust 1 month for every 2 hours. (At 9pm in May, use Apr; at 1am in May, use Jun.) See page 10 for more info.

Stars in Corvus

	LIGHT YEARS	MAGNITUDE	NAME ORIGIN
Gienah (γ)	185	2.6	Arabic for "wing"
Kraz (β)	140	2.6	Name of unknown origins
Minkar (ε)	303	3.0	Arabic for "beak"
Algorab (δ)	87	3.1	Arabic for "raven's wing"

Every star is ranked in terms of brightness, or magnitude. The brighter the star, the lower the magnitude. Stars with negative magnitudes are very bright. The sixth magnitude is the limit of human vision; the brightest stars are first magnitude or less.

Gienah (γ): The brightest star in the constellation, its name means "wing" in Arabic.

Kraz (β): The second-brightest star in the constellation. Strangely, its name is of unknown origins.

Minkar (ε): The star's name is of unknown origin. Some sources suggest it may derive from an Arabic word for "beak."

Algorab (δ): The dimmest of the constellation's four primary stars. Its name derives from Arabic for "raven's wing."

Mythology/History:

This constellation is said to represent Corvus the crow, Apollo's companion bird. The son of Zeus, Apollo was one of the most important gods of Olympus. This constellation was placed in the sky by Apollo to punish, rather than honor, his avian companion.

Apollo would often send Corvus on errands for him. One day, he asked the crow to fetch him a cup of water from a sacred spring to use in a ritual. Corvus flew off with the cup. On his way to the spring, he came upon a fig tree full of unripe fruit. He waited several days for the fruit to ripen, then ate his fill. Eventually, he arrived at the spring and filled the cup. Then, spotting a water snake in the stream, Corvus devised a plan to blame the creature for his tardiness. Corvus scooped the snake up and carried it, along with the cup, back to Apollo. When Corvus finally returned, he found that Apollo had used plain water for his ritual. Corvus explained that the snake blocked the stream and that he had just succeeded in overcoming the creature. Apollo saw through the crow's lie and condemned the crow to a lifetime of eternal thirst. This is why the crow has such a raspy voice.

Apollo cast Corvus, the cup and the snake into the sky for all to see. Corvus sits eternally above Hydra's tail, just out of reach of the constellation Crater, the cup filled with spring water.

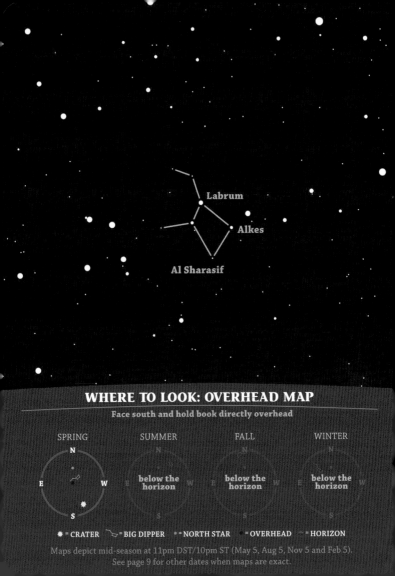

Labrum

Alkes

Al Sharasif

WHERE TO LOOK: OVERHEAD MAP

Face south and hold book directly overhead

SPRING   SUMMER   FALL   WINTER

N

E   W

S

below the horizon   below the horizon   below the horizon

✳ = CRATER   ⌐ = BIG DIPPER   ▪ = NORTH STAR   ● = OVERHEAD   ▪ = HORIZON

Maps depict mid-season at 11pm DST/10pm ST (May 5, Aug 5, Nov 5 and Feb 5).
See page 9 for other dates when maps are exact.

CRATER *(CRAY-ter)*

English Name: the cup

Size: small, 53rd largest

When to Look: most prominent in April and May and visible in the late evening sky from February through May

Notes: Tricky to find and difficult to trace, identifying this constellation is a good challenge for a dark spring night.

WHERE TO LOOK: HORIZON GRAPH

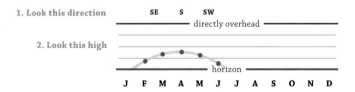

1. Look this direction   SE   S   SW

—————— directly overhead ——————

2. Look this high

———— horizon ————

J F M A M J J A S O N D

Shown 11pm DST/10pm ST on the 15th. If viewing earlier/later, adjust 1 month for every 2 hours. (At 9pm in May, use Apr; at 1am in May, use Jun.) See page 10 for more info.

77

Stars in Crater

	LIGHT YEARS	MAGNITUDE	NAME ORIGIN
Labrum (δ)	195	3.6	Latin reference to the Holy Grail
Alkes (α)	170	4.1	Arabic for "cup"
Al Sharasif (β)	266	4.5	Arabic for "ribs"

Every star is ranked in terms of brightness, or magnitude. The brighter the star, the lower the magnitude. Stars with negative magnitudes are very bright. The sixth magnitude is the limit of human vision; the brightest stars are first magnitude or less.

Labrum (δ): The brightest star in this very dim constellation. Its name comes from Latin and is a reference to the Holy Grail that Jesus drank from at the Last Supper.

Alkes (α): The alpha star of the constellation, though only the third-brightest star.

Al Sharasif (β): A dim star that lies along the back of Hydra, its name means "ribs" in Arabic, suggesting that it was originally considered to be part of that larger constellation.

Mythology/History:

Long identified as a cup, this constellation is associated with several different Greek myths. Most commonly, it is associated with the myth of the crow and the fig tree. The crow was the companion bird of the god Apollo. Apollo sent the crow on an errand to fetch water from a sacred spring for use in a ritual. On the way, the crow spotted a fig tree full of unripe fruit. It waited for the fruit to ripen and ate its fill. It then returned with the spring water, but it was too late, as Apollo had already used plain water in the ritual. To explain the delay, the crow blamed a water snake it found near the sacred spring, but Apollo saw through the lie and condemned the crow to a lifetime of eternal thirst.

In another legend, the constellation represents the wine cup of Icarius, who was shown the secret of wine-making by the god Dionysus. Icarius loaded his oxcart with wine and set out to share it with the shepherds of the region. They all drank too much and woke with terrible hangovers. Thinking Icarius had tried to poison them to steal their flocks, the shepherds murdered Icarius in his sleep. When they realized their error, the shepherds begged Dionysus for forgiveness. Dionysus placed Icarius's cup in the stars as a reminder of this wrongful death. Occasionally, this constellation is said to be the nectar cup of the gods of Olympus, which was filled by the shepherd boy Ganymede, who is immortalized in the constellation Aquarius (page 197).

Coma Star Cluster

Diadem

WHERE TO LOOK: OVERHEAD MAP

Face south and hold book directly overhead

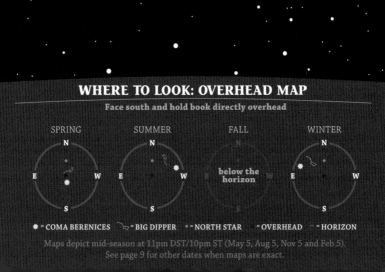

SPRING SUMMER FALL WINTER

below the horizon

✳ = COMA BERENICES ⌐◦ = BIG DIPPER •• = NORTH STAR — = OVERHEAD —— = HORIZON

Maps depict mid-season at 11pm DST/10pm ST (May 5, Aug 5, Nov 5 and Feb 5).
See page 9 for other dates when maps are exact.

COMA BERENICES *(COE-muh BER-uh-NICE-eez)*

English Name: Berenice's hair

Size: small, 42nd largest

When to Look: most prominent from March through June and visible in the late evening sky from February through July

Notes: A very dim and somewhat indistinct constellation, Coma Berenices is difficult to spot and tricky to trace. Under very good conditions, it actually does resemble a lock of hair. It lies in the direction of the north galactic pole and provides one of the clearest views in the sky out of our own galaxy.

WHERE TO LOOK: HORIZON GRAPH

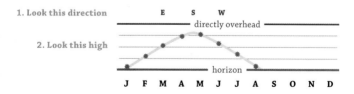

1. Look this direction

2. Look this high

Shown 11pm DST/10pm ST on the 15th. If viewing earlier/later, adjust 1 month for every 2 hours. (At 9pm in May, use Apr; at 1am in May, use Jun.) See page 10 for more info.

Stars in Coma Berenices

	LIGHT YEARS	MAGNITUDE	NAME ORIGIN
Diadem (α)	59	4.3	Greek for a type of simple crown

Every star is ranked in terms of brightness, or magnitude. The brighter the star, the lower the magnitude. Stars with negative magnitudes are very bright. The sixth magnitude is the limit of human vision; the brightest stars are first magnitude or less.

Diadem (α): The constellation's brightest star, with a magnitude of just 4.3.

The Coma Star Cluster: A loose open cluster of stars, sometimes called Melotte 111, which forms the tips of the strands of Berenice's hair. The brightest stars in the cluster are about fifth magnitude, making them just visible to the naked eye and easy to see in binoculars. The cluster covers an area about 5 degrees across (roughly half the width of your fist, held at arm's length).

North Galactic Pole: When we look toward Coma Berenices, we are looking perpendicular to the disk of our own galaxy. With relatively few stars and little interstellar dust in the way, we get one of our clearest views of deep space. The Coma Galaxy Cluster and the nearby Virgo Galaxy Cluster are in this region of space and each contains more than 1,000 galaxies that are hundreds of millions of light years from Earth. These galaxies are so far away you need a telescope to see them.

Mythology/History:

Coma Berenices is a modern constellation with ancient roots. Composed only of very dim stars, it was recognized by the ancient Greeks and described by Ptolemy but was considered to be part of the constellation Leo.

It has its own charming mythology, which dates back to antiquity, and thanks to Tycho Brahe's star catalog of 1602, it is now a constellation in its own right. This is one of the few constellations named after a historical figure.

The stars of this constellation are said to represent the hair of Queen Berenice II of Egypt, who ruled along with her husband King Ptolemy III Euergetes from 264–221 BC. In about 243 BC, the king set off on a dangerous military campaign in Syria to avenge the murder of his sister. Berenice, who was renowned for her long golden hair, vowed to sacrifice it to the gods if her husband returned victorious. When he did, she cut off her hair and placed it in the temple of Aphrodite. The next day, the lock of hair was missing, enraging the entire court. The quick-thinking court astronomer, Conon, explained that the goddess was so pleased she took it from the temple and placed it in the sky. He pointed to a group of faint stars northeast of the constellation Leo as the new home of Berenice's hair.

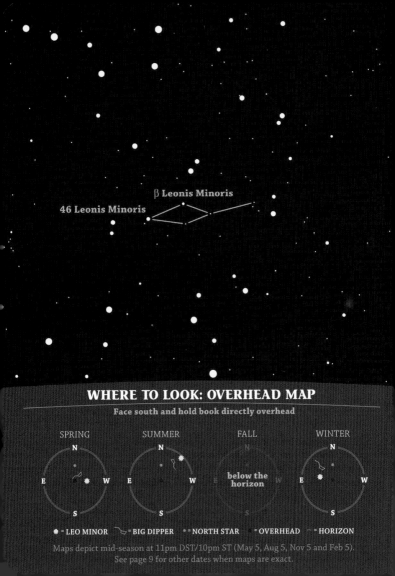

β Leonis Minoris

46 Leonis Minoris

WHERE TO LOOK: OVERHEAD MAP

Face south and hold book directly overhead

SPRING

SUMMER

FALL

below the horizon

WINTER

✳ = LEO MINOR ⌇ = BIG DIPPER • = NORTH STAR ✱ = OVERHEAD ⌒ = HORIZON

Maps depict mid-season at 11pm DST/10pm ST (May 5, Aug 5, Nov 5 and Feb 5).
See page 9 for other dates when maps are exact.

LEO MINOR (LEE-oh MY-ner)

English Name: the little lion

Size: small, 64th largest

When to Look: most prominent from February through May and visible in the late evening sky from December through June

Notes: A small, dim, modern constellation lying in an undistinguished region of the sky. It contains just two fourth-magnitude stars, and none of its stars have a common name. It is relatively easy to locate because it lies between the large, bright constellations of Leo and Ursa Major (which contains the Big Dipper), but its stars are hard to spot and its shape is undistinguished.

WHERE TO LOOK: HORIZON GRAPH

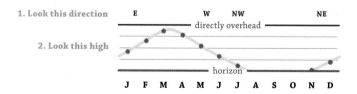

1. Look this direction

2. Look this high

Shown 11pm DST/10pm ST on the 15th. If viewing earlier/later, adjust 1 month for every 2 hours. (At 9pm in May, use Apr; at 1am in May, use Jun.) See page 10 for more info.

Stars in Leo Minor

	LIGHT YEARS	MAGNITUDE	NAME ORIGIN
46 Leonis Minoris	98	3.8	No common name
β **Leonis Minoris**	100	4.2	No common name

Every star is ranked in terms of brightness, or magnitude. The brighter the star, the lower the magnitude. Stars with negative magnitudes are very bright. The sixth magnitude is the limit of human vision; the brightest stars are first magnitude or less.

This small constellation contains just two fourth-magnitude stars, neither of which has a common name nor much of interest for the stargazer or even an amateur with a good-sized telescope.

46 Leonis Minoris: With a magnitude of 3.8, it is the brightest star in the constellation. It carries only a Flamsteed designation and has no Greek letter nor common name.

β **Leonis Minoris**: The second-brightest star in the constellation, and the only one to still carry a Greek letter Bayer designation.

Mythology/History:

Leo Minor is a modern constellation created by Polish astronomer Johannes Hevelius in the late seventeenth century to fill one of the areas that the ancient Greeks referred to as *amorphotoi*, meaning "unformed" or "shapeless." Hevelius said that he named these stars "the Little Lion" because they were "of the same nature" as their neighbors in Leo (the lion) and Ursa Major (the great bear).

The constellation is one of seven that Hevelius contributed to our modern maps of the sky. The other six are Canes Venatici, Lacerta, Lynx, Scutum, Sextans and Vulpecula.

This constellation has the unusual distinction of being the only one in the Northern Hemisphere that does not have an α (alpha) star. Hevelius himself never gave his stars Bayer designations; in fact, Francis Baily assigned Greek letters to the stars 150 years later.

Apparently by accident, Baily did not label the brightest star in Leo Minor when he published his star catalog of 1845. Baily labeled the second-brightest star in the constellation β Leonis Minoris, but did not designate an α. Today, the constellation's brightest star carries only its Flamsteed designation of 46 Leonis Minoris.

α **Antliae**

ε **Antliae**

WHERE TO LOOK: OVERHEAD MAP

Face south and hold book directly overhead

SPRING

SUMMER

FALL

WINTER

below the horizon

below the horizon

✷ = ANTLIA   ⌐◝ = BIG DIPPER   •• = NORTH STAR   ▫ = OVERHEAD   ▬ = HORIZON

Maps depict mid-season at 11pm DST/10pm ST (May 5, Aug 5, Nov 5 and Feb 5).
See page 9 for other dates when maps are exact.

ANTLIA *(ANT-lee-uh)*

English Name: the air pump

Size: small, 62nd largest

When to Look: visible in the late evening sky in March and April

Notes: A small, dim, modern constellation. It is difficult to spot and hard to trace. Its brightest stars are of the fourth magnitude and none of its stars have common names. Under good conditions, it is fully visible throughout the United States, but its faint stars are hard to pick out close to the horizon.

WHERE TO LOOK: HORIZON GRAPH

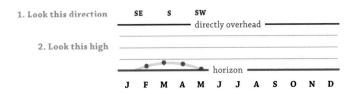

1. Look this direction

SE   S   SW
directly overhead

2. Look this high

horizon

J F M A M J J A S O N D

Shown 11pm DST/10pm ST on the 15th. If viewing earlier/later, adjust 1 month for every 2 hours. (At 9pm in Apr, use Mar; at 1am in Apr, use May.) See page 10 for more info.

Stars in Antlia

	LIGHT YEARS	MAGNITUDE	NAME ORIGIN
α **Antliae**	365	4.3	No common name
ε **Antliae**	700	4.5	No common name

Every star is ranked in terms of brightness, or magnitude. The brighter the star, the lower the magnitude. Stars with negative magnitudes are very bright. The sixth magnitude is the limit of human vision; the brightest stars are first magnitude or less.

None of the stars in Antlia have common names and most are near the limit of naked-eye visibility.

α **Antliae**: It shines with a magnitude of 4.3 from 275 light years away. It has no common name.

ε **Antliae**: The constellation's second-brightest star at magnitude 4.5, it lies 315 light years from Earth.

Mythology/History:

This constellation was created by French astronomer Nicolas Louis de Lacaille in the mid-eighteenth century in commemoration of the air pump. Like most of the 14 constellations that de Lacaille defined, it is very faint, named after an important scientific instrument and looks nothing like its namesake.

The original name for the constellation was *Antlia Pneumatica*, which means "Air Pump" in Latinized Greek. Although the constellation is still called the "Air Pump" in English, the International Astronomical Union short-ened its official name to Antlia.

De Lacaille named this constellation to honor the work of French physicist Denis Papin. In the late 1670s, Papin studied in London with Robert Boyle, the renowned seventeenth-century theologian, scientist and inventor. Boyle is best known for his work on the air pump and his study of the properties of gases.

Papin built on Boyle's work, but sadly, Papin received little credit for many of his ideas. He fell into poverty, died destitute and many of his discoveries were credited to others.

THE SUMMER SKY
(overhead, facing south)

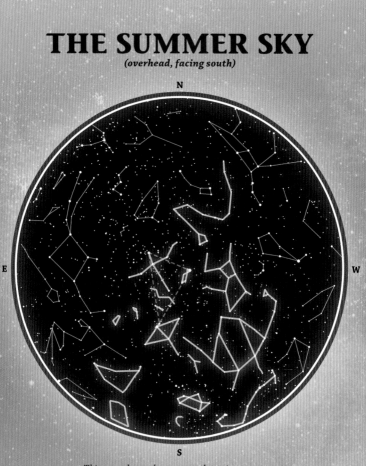

N

E

W

S

This map shows the summer sky as it appears at
1am on Jul 5, 11pm on Aug 5 and 9pm on Sep 5.
For other times this chart can be used, see page 9.

IN THE SUMMER MONTHS, the dark side of the Earth faces toward the center of our galaxy. The evenings are warm, and on clear, dark nights the rich, glowing band of the Milky Way contains endless wonders. It is easy to spend many hours gazing into the rich star fields of our galaxy. Summer is also when we have some of our strongest meteor activity—adding the unpredictable treat of a "shooting star" to a summer evening spent watching the night sky.

The most prominent feature in the sky this time of year is the Summer Triangle (highlighted at left with a gray dotted line)—a trio of bright stars from the constellations Aquila (page 103), Cygnus (page 95) and Lyra (page 99). Stretching through the middle of the Summer Triangle is the soft, glowing band of the Milky Way.

Low in the southern sky, you can see the constellations Sagittarius (page 111) and Scorpius (page 107). When you see Sagittarius, you are looking toward the center of our own galaxy. With all these bright stars and constellations, don't overlook the delicate beauty of the Milky Way itself, and be sure not to miss the tiny, beautiful constellation Delphinus (page 139)!

The summer constellations are in blue on the seasonal sky map to the left.

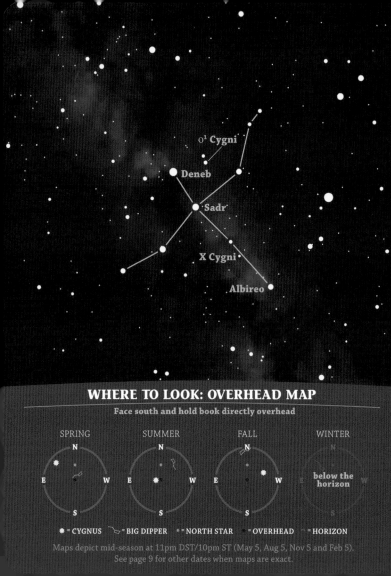

O¹ Cygni

Deneb

Sadr

X Cygni

Albireo

WHERE TO LOOK: OVERHEAD MAP

Face south and hold book directly overhead

SPRING

SUMMER

FALL

WINTER

below the
horizon

✳ = CYGNUS ⌐ = BIG DIPPER • = NORTH STAR = OVERHEAD ⌐ = HORIZON

Maps depict mid-season at 11pm DST/10pm ST (May 5, Aug 5, Nov 5 and Feb 5).
See page 9 for other dates when maps are exact.

CYGNUS *(SIG-nus)*

English Name: the swan

Size: large, 16th largest

When to Look: most prominent from July through October and visible in the late evening sky from May through December

Notes: A beautiful constellation that lies across a rich section of the Milky Way. Easy to spot because of its bright star Deneb, which marks the swan's tail and also forms the northeast corner of the Summer Triangle. The main stars of the constellation form a cross, making it easy to trace and leading to its popular nickname, the Northern Cross.

WHERE TO LOOK: HORIZON GRAPH

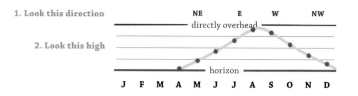

1. Look this direction

NE   E   W   NW

directly overhead

2. Look this high

horizon

J F M A M J J A S O N D

Shown 11pm DST/10pm ST on the 15th. If viewing earlier/later, adjust 1 month for every 2 hours. (At 9pm in Jun, use May; at 1am in Jun, use Jul.) See page 10 for more info.

95

Stars in Cygnus

	LIGHT YEARS	MAGNITUDE	NAME ORIGIN
Deneb (α)	1,550	1.3	Arabic for "tail"
Sadr (γ)	1500	2.2	Arabic for "breast"
Albireo (β)	390	3.1	Result of translation errors
X **Cygni**	350	6.8*	No common name
o¹ **Cygni**	600	3.8	No common name

Every star is ranked in terms of brightness, or magnitude. The brighter the star, the lower the magnitude. Stars with negative magnitudes are very bright. The 6th magnitude is the limit of human vision; the brightest stars are first-magnitude or less.

* Varies from 3.3 to 14.0

Deneb (α): The 14th-brightest star in our sky, Deneb is one of the brightest stars in the galaxy and the most distant of all the first-magnitude stars. Deneb has a diameter larger than the orbit of the Earth.

Sadr (γ): The constellation's second-brightest star, it marks the point where the swan's wings meet the body.

X **Cygni**: A Mira-class variable star (page 186) that lies midway along the swan's neck.

Albireo (β): The head of the swan. What appears to the naked eye as a single star is actually a binary star system. A telescope or a very steady hand with binoculars reveals a bright-golden primary star with a brilliant sapphire-blue companion.

o¹ **Cygni**: One of the fainter stars in the constellation visible to the naked eye. Binoculars reveal it to be a multiple star system and one of the most beautiful in the sky.

The Milky Way passes through the middle of this constellation, making it a very interesting area to scan with binoculars.

Mythology/History:

Civilizations all over the world have long depicted this constellation as a bird. In the earliest Greek literature, it is referred to simply as a bird, but by the time of Ptolemy, the constellation was recognized as a swan and associated with Zeus's seduction of Leda, the Queen of Sparta.

According to this legend, Zeus became enamored with Leda, but she remained loyal to her husband, King Tyndareus. To deceive Leda, Zeus transformed himself into a swan and approached her. Leda, suspecting nothing, allowed the swan to come and rest on her lap as she rested. When she fell asleep, Zeus consorted with her. Zeus then flew off to the heavens, still in the guise of a swan, and placed the image of the swan among the stars.

From this union, Leda produced two eggs. From one egg hatched Helen of Troy. The twins Pollux and Castor hatched from the other egg; they are immortalized in the constellation Gemini. It was said that the shell of this second egg was displayed at a temple in Sparta, hung from the ceiling with ribbons.

Today, the stars of Cygnus are popularly referred to as the Northern Cross because of their similarity to the Christian symbol.

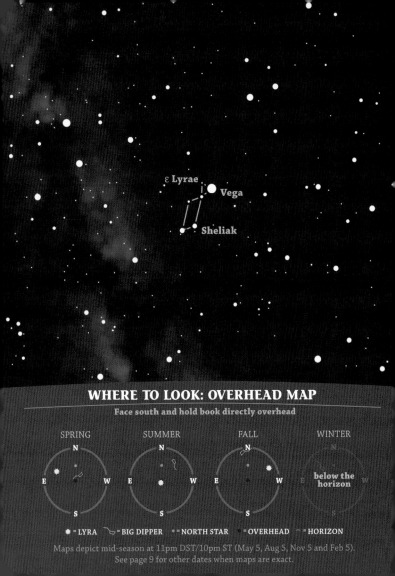

ε **Lyrae**

Vega

Sheliak

WHERE TO LOOK: OVERHEAD MAP

Face south and hold book directly overhead

SPRING

SUMMER

FALL

WINTER

below the horizon

✳ = LYRA   ↝ = BIG DIPPER   • = NORTH STAR   • = OVERHEAD   = HORIZON

Maps depict mid-season at 11pm DST/10pm ST (May 5, Aug 5, Nov 5 and Feb 5).
See page 9 for other dates when maps are exact.

LYRA *(LYE-ruh)*

English Name: the lyre

Size: small, 52nd largest

When to Look: most prominent from June through September and visible in the late evening sky from April through November

Notes: One of the gems of the northern sky, Lyra is clearly marked by its brilliant star Vega, which marks the north-west corner of the Summer Triangle. The constellation also contains a number of multiple star systems of great interest to astronomers.

WHERE TO LOOK: HORIZON GRAPH

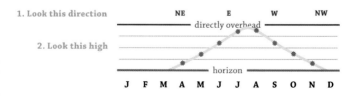

Shown 11pm DST/10pm ST on the 15th. If viewing earlier/later, adjust 1 month for every 2 hours. (At 9pm in Jun, use May; at 1am in Jun, use Jul.) See page 10 for more info.

Stars in Lyra

	LIGHT YEARS	MAGNITUDE	NAME ORIGIN
Vega (α)	25	0.0	Arabic for "vulture"
Sheliak (β)	900	3.85*	Arabic for "harp"
ε **Lyrae**	162	3.9	No common name

Every star is ranked in terms of brightness, or magnitude. The brighter the star, the lower the magnitude. Stars with negative magnitudes are very bright. The sixth magnitude is the limit of human vision; the brightest stars are first magnitude or less.

* Varies from 3.4 to 4.3

Vega (α): The third-brightest star in our sky, its brilliant blue-white color is easy to see. Although Vega currently sits nearly 40 degrees from the celestial pole, it lies along the path traced by the Earth's axis as it slowly wobbles over the centuries. During the last ice age, the Earth's axis pointed toward Vega, and this brilliant gem was the North Star. Vega will become the North Star again in about 12,500 years.

Sheliak (β): What looks like a single star, even through telescopes, is actually a very close pair of stars which orbit each other in just 13 days. The stars are so close together that they practically touch. As they orbit, they eclipse each other, causing the magnitude of the system to vary between 3.4 and 4.3.

ε **Lyrae**: Another interesting multiple star, known as the "double-double." To the naked eye, it appears to be a single star of fourth magnitude. Binoculars, or exceptional eyesight, reveal two stars with magnitudes of 4.6 and 4.7. A medium-sized telescope makes it clear that these are double stars as well.

The Milky Way passes through the southeast corner of the constellation.

Mythology/History:

An ancient constellation, Lyra dates back to Babylonian times. According to myth, Hermes created the lyre using a tortoise shell and the horns and gut of cattle he stole from the god Apollo. When Apollo came to retrieve his cattle, he was enchanted by the music that Hermes was making and let him keep the cattle in exchange for the lyre. Apollo gave the lyre to Orpheus, the son of Calliope, the muse of epic poetry. Orpheus played the lyre well and was held in great esteem, even by the gods. It was said his songs were so beautiful that he could charm rocks, trees and wild animals. While Orpheus was still a young man, his beloved wife, Eurydice, died.

In one of the most famous stories in mythology, he descended into the underworld to find her, singing and playing as he went. Even Hades, god of the underworld, was charmed and agreed to release Eurydice. He gave just one condition—that Orpheus not look back as they ascended to the world of the living. Nevertheless, a joyful Orpheus momentarily forgot and glanced over his shoulder. As he did, his wife faded into dust.

Overcome with grief, Orpheus spent the rest of his days singing beautiful, sad songs. Many women wanted to marry him, but he rebuffed them. One day, a group of maidens became enraged that he wouldn't choose one of them, and they murdered him. Zeus reunited him with his wife in the underworld, placing the lyre in the heavens.

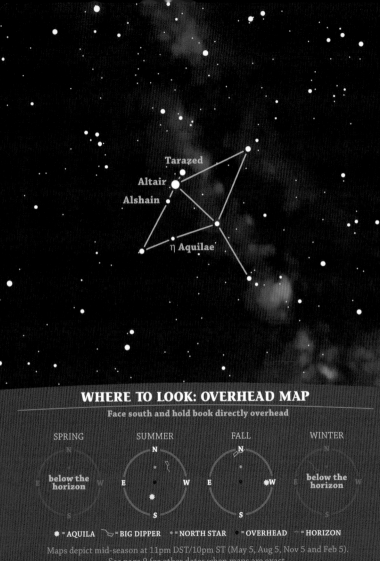

Tarazed

Altair

Alshain

η Aquilae

WHERE TO LOOK: OVERHEAD MAP

Face south and hold book directly overhead

SPRING	SUMMER	FALL	WINTER
below the horizon			below the horizon

* = AQUILA   = BIG DIPPER   •• = NORTH STAR   • = OVERHEAD   — = HORIZON

Maps depict mid-season at 11pm DST/10pm ST (May 5, Aug 5, Nov 5 and Feb 5).
See page 9 for other dates when maps are exact.

AQUILA *(ACK-will-uh)*

English Name: the eagle

Size: medium, 22nd largest

When to Look: most prominent in August and September and visible in the late evening sky from June through October

Notes: Aquila is easy to spot due to its bright star Altair, which forms the southern point of the Summer Triangle. The other two corners are marked by Deneb in Cygnus and Vega in Lyra. The rest of the constellation is much dimmer, but its clear shape makes it relatively easy to trace under reasonably dark skies.

WHERE TO LOOK: HORIZON GRAPH

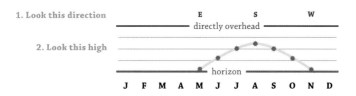

Shown 11pm DST/10pm ST on the 15th. If viewing earlier/later, adjust 1 month for every 2 hours. (At 9pm in Jun, use May; at 1am in Jun, use Jul.) See page 10 for more info.

Stars in Aquila

	LIGHT YEARS	MAGNITUDE	NAME ORIGIN
Altair (α)	17	0.8	Arabic for abbreviation of "the flying eagle"
Tarazed (γ)	460	2.7	Persian for "the beam of the scale"
Alshain (β)	48	3.7	Arabic for the entire constellation, Persian derivative of "falcon"
η **Aquilae**	1,450	4.1*	No common name

Every star is ranked in terms of brightness, or magnitude. The brighter the star, the lower the magnitude. Stars with negative magnitudes are very bright. The sixth magnitude is the limit of human vision; the brightest stars are first magnitude or less.

* Varies from 4.5 to 3.7 every 7 days, 4 hours.

Altair (α): Altair is the eighth-brightest star in our sky and one of our closer neighbors, just 17 light years from Earth. It marks the southern point of the Summer Triangle.

Tarazed (γ): The constellation's second-brightest star. Its name refers to the asterism it forms together with Altair and Alshain—a line of three stars sometimes thought to resemble Orion's Belt.

Alshain (β): Its magnitude of 3.7 equals that of the nearby variable star Eta Aquilae when it is at its brightest.

η **Aquilae**: A bright "Cepheid"-type variable star that ranges in brightness from magnitude 4.5 to 3.7, a cycle which takes 7 days, 4 hours. It is easily compared to nearby Alshain.

Antinous: The faint group of stars below the main figure of the eagle was once a separate constellation known as Antinous. The constellation was created by Roman Emperor Hadrian in 132 AD in honor of a beloved boy in his court who drowned while on a trip up the Nile. These stars are also said to represent the boy Ganymede being carried to Olympus by Aquila.

Mythology/History:

Recognized as a bird since at least 1200 BC, Aquila is an ancient constellation. In Greek mythology, Aquila is the eagle of Zeus, king of the gods. The eagle was considered a royal bird by the Greeks, much as the lion is considered a royal animal today.

According to legend, when Zeus was coming of age, he took it upon himself to overthrow the ancient Titans— giants who reigned over the Earth. As he was preparing to rally the gods of Olympus for battle, Aquila appeared to him. Zeus took this as an omen and accepted Aquila as his own. When Zeus and the gods of Olympus defeated the Titans, Zeus honored Aquila by placing it among the stars.

Aquila remained Zeus's trusted servant and carried Zeus's famous thunderbolts. Zeus also sent Aquila on special errands—the most famous of these was the abduction of the shepherd boy Ganymede. Ganymede was a young shepherd boy considered to be the most beautiful child ever born to mortals. Zeus took the boy to Mount Olympus to serve as cupbearer to the gods. It was Aquila that carried Ganymede to Mount Olympus. Ganymede is also immortalized among the stars; he is represented in the constellation Aquarius, just south and east of Aquila.

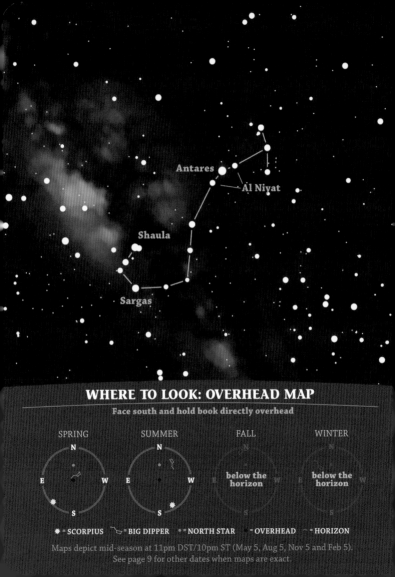

Antares

Al Niyat

Shaula

Sargas

WHERE TO LOOK: OVERHEAD MAP
Face south and hold book directly overhead

SPRING

SUMMER

FALL

below the horizon

WINTER

below the horizon

✳ = SCORPIUS   ⌐⌐ = BIG DIPPER   • = NORTH STAR   ◆ = OVERHEAD   ⌒ = HORIZON

Maps depict mid-season at 11pm DST/10pm ST (May 5, Aug 5, Nov 5 and Feb 5).
See page 9 for other dates when maps are exact.

SCORPIUS *(SCOR-pee-us)*

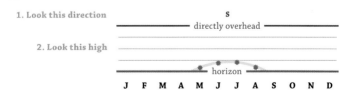

zodiac

English Name: the scorpion

Size: medium, 33rd largest

When to Look: visible in the late evening sky from June through July

Notes: A well-known constellation and one of the few that actually looks like its namesake. Its bright star, Antares, shines red and is located in the anatomically correct location for the scorpion's heart. Easy to locate and trace, it is only above the horizon for a few hours and can often be masked by ground level haze or obstructions. Even under ideal conditions it may be partially hidden above 40 degrees north latitude.

WHERE TO LOOK: HORIZON GRAPH

1. Look this direction

S

directly overhead

2. Look this high

horizon

J F M A M J J A S O N D

Shown 11pm DST/10pm ST on the 15th. If viewing earlier/later, adjust 1 month for every 2 hours. (At 9pm in Jun, use May; at 1am in Jun, use Jul.) See page 10 for more info.

Stars in Scorpius

	LIGHT YEARS	MAGNITUDE	NAME ORIGIN
Antares (α)	520	1.0	Arabic for "rival of Mars"
Shaula (λ)	365	1.6	Arabic for "the stinger"
Sargas (θ)	270	1.9	Sumerian origins; name unknown
Al Niyat (τ & σ)	430 734	2.8 2.9	Arabic for "arteries"

Every star is ranked in terms of brightness, or magnitude. The brighter the star, the lower the magnitude. Stars with negative magnitudes are very bright. The sixth magnitude is the limit of human vision; the brightest stars are first magnitude or less.

Antares (α): This brilliant red star sits in the anatomically correct location for the scorpion's heart. Its common name comes from Greek and means "rival of Mars." Occasionally Mars passes very close to Antares, each shining brilliant red. In 2016, Mars will spend all summer near Antares, passing closest in August. One of the largest stars known, it is 700 times the diameter of our sun.

Shaula (λ): The second-brightest star in the constellation, and one of the brightest second-magnitude stars in the sky.

Sargas (θ): The third-brightest star in the constellation.

Al Niyat (τ & σ): This name has been given to both of the two stars that sit just above and below Antares.

Ecliptic: The sun enters Scorpius on November 21 and departs just a week later on November 28. Its astrological period runs October 23 to November 21.

The Milky Way passes though Scorpius.

Mythology/History:

In Greek mythology, this constellation honors the tiny scorpion that killed Orion with its sting. Orion was said to be a mighty hunter, but also very boastful—a failing never much tolerated by the gods. According to legend, he once claimed to be such a great hunter that no beast could be a match for him. Hera, the wife of Zeus, sent a tiny scorpion to teach him a lesson. Orion killed the scorpion with his club, but only after the tiny creature had dealt him a fatal sting. The scorpion was placed in the sky to honor its sacrifice and as a reminder to others who might be as boastful as Orion. Scorpius was placed on the opposite side of the sky from Orion, always chasing after him, but never catching him. As Scorpius is rising, Orion sets and the two are never seen in the sky at the same time.

Scorpius was once a much larger constellation. Up until about the time of Ptolemy, it included the stars that we now call Libra. Several of the stars of Libra still carry the names of the scorpion's claws as a reminder of their history (page 65).

In western astrology, the constellation is known as Scorpio. The astrological symbol for Scorpio (♏) is thought to be a combination of an ancient symbol for medicine combined with the scorpion's stinger and is said to represent both life and death.

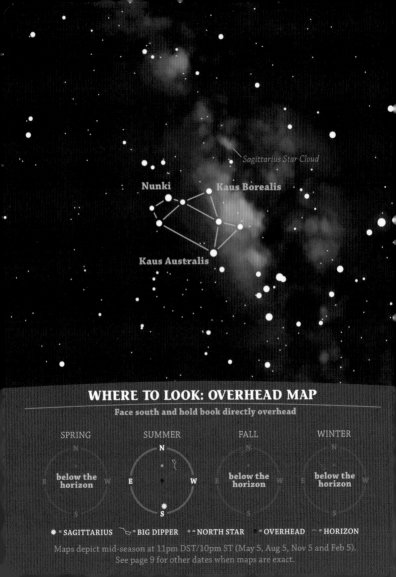

Sagittarius Star Cloud

Nunki

Kaus Borealis

Kaus Australis

WHERE TO LOOK: OVERHEAD MAP

Face south and hold book directly overhead

SPRING SUMMER FALL WINTER

N N N N

below the horizon E W **below the horizon** **below the horizon**

S S

✹ = SAGITTARIUS   ⌐⌐ = BIG DIPPER   • = NORTH STAR   ✷ = OVERHEAD   ⌒ = HORIZON

Maps depict mid-season at 11pm DST/10pm ST (May 5, Aug 5, Nov 5 and Feb 5).
See page 9 for other dates when maps are exact.

SAGITTARIUS *(SAJ-ih-TARE-ee-us)*

zodiac

English Name: the archer

Size: large, 14th largest

When to Look: visible in the late evening sky from June through August

Notes: Classical depictions show these stars forming a mythical centaur firing a bow. Many modern charts, however, show the bright stars of "the archer" forming a teapot instead. While much of the constellation is dim and difficult to see from of our region, the teapot is easy to spot and easy to trace. The teapot also lies in the direction of our galactic center. When you look at Sagittarius you are looking toward the heart of the Milky Way.

WHERE TO LOOK: HORIZON GRAPH

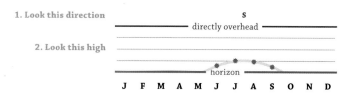

1. Look this direction

2. Look this high

S

directly overhead

horizon

J F M A M J J A S O N D

Shown 11pm DST/10pm ST on the 15th. If viewing earlier/later, adjust 1 month for every 2 hours. (At 9pm in Jul, use Jun; at 1am in Jul, use Aug.) See page 10 for more info.

Stars in Sagittarius

	LIGHT YEARS	MAGNITUDE	NAME ORIGIN
Kaus Australis (ε)	145	1.8	Arabic for "bow" and Latin for "southern"
Nunki (σ)	225	2.02	Babylonian for "Herald of the Sea"
Kaus Borealis (λ)	77	2.8	Arabic for "bow" and Latin for "northern"

Every star is ranked in terms of brightness, or magnitude. The brighter the star, the lower the magnitude. Stars with negative magnitudes are very bright. The sixth magnitude is the limit of human vision; the brightest stars are first magnitude or less.

Kaus Australis (ε): It marks the southern tip of the archer's bow, and the southeast corner of the teapot.

Nunki (σ): The second-brightest star in the constellation, it bears an ancient Babylonian proper name. The Babylonians knew this star as "Herald of the Sea," because it rose just before the constellations they associated with water.

Kaus Borealis (λ): Marks the northern tip of the bow.

Sagittarius Star Cloud (M24): Visible to the naked eye as a dim, cloud-like region of the Milky Way, it provides an unobstructed view for 16,000 light years. Binoculars reveal the richest star field in the sky with about 1,000 stars in a single field of view.

Ecliptic: The sun is located in Sagittarius from December 16 through January 17. Its astrological period runs November 22 to December 21.

Mythology/History:

An ancient constellation dating back to the Babylonians, the Greeks were the first to identify these stars as an archer. The constellation is often depicted as a centaur—a mythical creature with a horse's body and legs, and the trunk, arms and head of a man. However, none of the centaurs in Greek mythology ever used a bow. Classical authorities identify this constellation instead with the satyr Crotus.

In classical Greek mythology, satyrs were sons of the god Pan and shared his distinctive appearance—they had the body of a man and the legs and horns of a goat. Crotus grew up in the company of the Muses, the nine nymphs considered to be the source of inspiration in the arts and sciences. Living an inspired life, Crotus became a highly accomplished hunter, athlete, musician and artist. It is said that Crotus invented rhythm in music by clapping his hands to mark time to the Muses' lyrical songs. Some ancient authorities also claim that he invented archery.

When Crotus died, the Muses asked their father, Zeus, to honor their childhood companion for his great skill and his contributions to the arts. At their request, Zeus placed Crotus among the stars.

The astrological symbol for Sagittarius (♐) is a stylized bow and arrow.

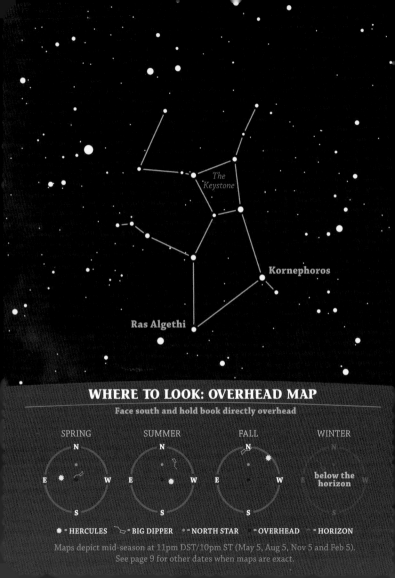

The Keystone

Kornephoros

Ras Algethi

WHERE TO LOOK: OVERHEAD MAP
Face south and hold book directly overhead

SPRING   SUMMER   FALL   WINTER

below the horizon

✳ = HERCULES   ⌐◡ = BIG DIPPER   • = NORTH STAR   ▪ = OVERHEAD   ▭ = HORIZON

Maps depict mid-season at 11pm DST/10pm ST (May 5, Aug 5, Nov 5 and Feb 5).
See page 9 for other dates when maps are exact.

HERCULES *(HER-kyuh-leez)*

English Name: the hero, Heracles

Size: very large, 5th largest

When to Look: most prominent from May through August and visible in the late evening sky from April through October

Notes: One of the largest constellations in the sky, it contains no truly bright stars. With a good bit of imagination, you can make out the shape of a kneeling man here, though he is upside down. The most distinctive group of stars here is the asterism known as the Keystone, found near the center of the constellation.

WHERE TO LOOK: HORIZON GRAPH

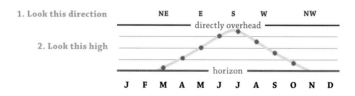

1. Look this direction

NE E S W NW

directly overhead

2. Look this high

horizon

J F M A M J J A S O N D

Shown 11pm DST/10pm ST on the 15th. If viewing earlier/later, adjust 1 month for every 2 hours. (At 9pm in Jun, use May; at 1am in Jun, use Jul.) See page 10 for more info.

Stars in Hercules

	LIGHT YEARS	MAGNITUDE	NAME ORIGIN
Kornephoros (β)	100	2.8	Greek for "club bearer"
Ras Algethi (α)	220	2.8	Arabic for "kneeler's head"

Every star is ranked in terms of brightness, or magnitude. The brighter the star, the lower the magnitude. Stars with negative magnitudes are very bright. The sixth magnitude is the limit of human vision; the brightest stars are first magnitude or less.

Kornephoros (β): The brightest star in the constellation. In 1899, astronomer W. W. Campbell discovered that it was actually a two star system. The stars are too close together to split with a telescope, but each star gives off distinctive wavelengths of light. Double stars that are resolved by their light spectrum are called spectroscopic binaries.

Ras Algethi (α): Ras Algethi is one of the largest stars known, measuring about 600 times the diameter of our sun—a diameter larger than the orbit of Mars. It is also a variable star, fluctuating in size and brightness.

The Keystone: It is a distinctive asterism that serves as a good reference point for tracing the rest of the constellation.

The Milky Way touches the far eastern edge of the constellation.

Mythology/History:

It's a bit difficult to see the figure of a person here, as he's upside down and kneeling. Despite this unusual posture, he clearly has a narrow waist and broad shoulders, and this constellation has been associated with heroic strongmen since Babylonian times.

In Roman mythology, these stars are associated with the hero Hercules, the half-mortal son of Jupiter. There are many stories about Hercules, but the Twelve Labors of Hercules are the most famous—Hercules was indentured to King Eurystheus and completed twelve labors to earn his freedom: kill the Nemean Lion; slay the Lernean Hydra; capture the Ceryneian Doe; capture the Erymanthian Boar; clean the Augean Stables in a day; drive away the Stymphalian Birds; capture the Cretan Bull; round up the Mares of Diomedes; obtain the magical girdle from the Amazonian Queen, Hippolyta; herd the cattle of Geryon; steal the apples of Hesperides; and capture Cerberus, the three-headed dog who guarded the gates of hell.

Many constellations are associated with Hercules, including Draco, which represents the dragon that guarded the apples of Hesperides which originally belonged to Hera (page 121). In one telling, Hercules shot Draco with a poisoned arrow. In the sky, Hercules' foot is atop Draco's head, a symbol of his triumph.

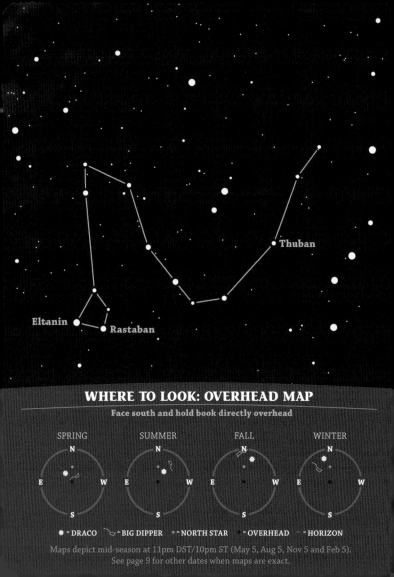

Eltanin

Rastaban

Thuban

WHERE TO LOOK: OVERHEAD MAP

Face south and hold book directly overhead

SPRING　　　　SUMMER　　　　FALL　　　　WINTER

✳ = DRACO ～ = BIG DIPPER • = NORTH STAR • = OVERHEAD ◡ = HORIZON

Maps depict mid-season at 11pm DST/10pm ST (May 5, Aug 5, Nov 5 and Feb 5).
See page 9 for other dates when maps are exact.

DRACO *(DRAY-co)*

English Name: the dragon

Size: very large, 8th largest

When to Look: most prominent from April through August and visible in the late evening sky throughout the year

Notes: A large, beautiful, but relatively dim constellation high in the northern skies. Its long, snaking shape wraps around Ursa Minor (the Little Dipper) on three sides. Fairly easy to locate, it requires darker skies to trace. Circumpolar above 44 degrees north latitude, it is visible for at least part of every night of the year throughout most of the United States.

WHERE TO LOOK: HORIZON GRAPH

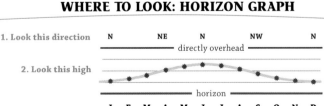

Shown 11pm DST/10pm ST on the 15th. If viewing earlier/later, adjust 1 month for every 2 hours. (At 9pm in Jun, use May; at 1am in Jun, use Jul.) See page 10 for more info.

Stars in Draco

	LIGHT YEARS	MAGNITUDE	NAME ORIGIN
Eltanin (γ)	148	2.2	Arabic for "serpent" and refers to entire constellation
Rastaban (β)	362	2.8	Arabic for "dragon's head"
Thuban (α)	310	3.7	Arabic for "dragon"

Every star is ranked in terms of brightness, or magnitude. The brighter the star, the lower the magnitude. Stars with negative magnitudes are very bright. The sixth magnitude is the limit of human vision; the brightest stars are first magnitude or less.

Eltanin (γ): The brightest star in the constellation, with a magnitude of 2.2. Its name means "serpent" and is the Arabic name for the entire constellation. The star passes through the zenith (straight overhead) in London, England, and was intensely studied by eighteenth-century astronomers.

Rastaban (β): The constellation's second-brightest star, with a magnitude of 2.8. Its name is Arabic for "dragon's head."

Thuban (α): Now an unremarkable star of magnitude 3.7, this star was well known in ancient Egypt. At that time, 5,000 years ago, the Earth's axis pointed directly at Thuban, which served as the North Star. Egyptian temples and The Great Pyramids were built to face toward Thuban. Due to the effects of procession (the slow wobble of the Earth's axis), the Earth's axis now points toward Polaris and Thuban lies more than 25 degrees away from the pole.

Mythology/History:

The ancient Egyptians identified these stars with the goddess Tawaret, who was depicted as a ferocious deity with the body of a crocodile.

This ancient constellation is also associated with nearly every dragon in Greek mythology. Often these dragons were said to be sleepless and ever-vigilant, making the circumpolar stars of the constellation a natural fit. Two of these stories are of particular interest.

In one legend, this constellation represents Tiamat, one of the ancient Titans who battled against the gods of Olympus. In the battle, Tiamat transformed herself into a huge dragon. The warrior goddess Athena defeated Tiamat and flung her body into the heavens.

In another legend, the constellation represents Ladon, a sleepless dragon with 100 heads and the guardian of a tree that belonged to Hera. The tree produced golden apples, which gave their owner immortality. The eleventh of the twelve labors of Hercules was to steal the golden apples from Hera's garden. Hercules slew Ladon with a poisoned arrow and made off with several apples. Saddened by the loss of her guardian, Hera placed Ladon in the stars. When Zeus placed Hercules in the heavens, he placed the hero's foot on top of Ladon's head, as a reminder of the struggle and of Hercules' triumph.

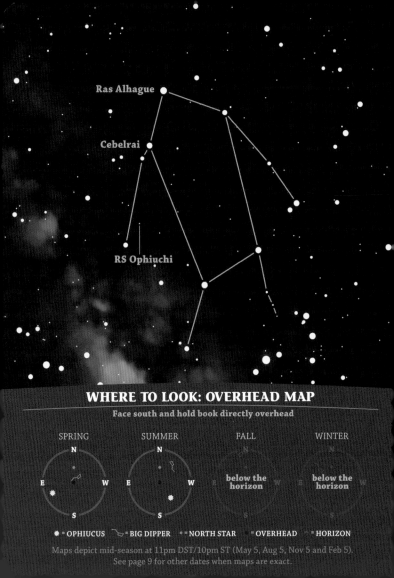

Ras Alhague

Cebelrai

RS Ophiuchi

WHERE TO LOOK: OVERHEAD MAP

Face south and hold book directly overhead

SPRING

SUMMER

FALL
below the horizon

WINTER
below the horizon

☀ = OPHIUCUS   = BIG DIPPER   = NORTH STAR   = OVERHEAD   = HORIZON

Maps depict mid-season at 11pm DST/10pm ST (May 5, Aug 5, Nov 5 and Feb 5).
See page 9 for other dates when maps are exact.

OPHIUCUS (OFF-ee-YOO-kus)

English Name: the serpent holder

Size: large, 11th largest

When to Look: most prominent in July and visible in the late evening sky from May through September

Notes: A large, diffuse constellation made up mostly of dim stars. It is difficult to trace, but with some effort it is possible to pick out the rough figure of a person here when the skies are dark. It occupies an interesting region of the sky, crossing the celestial equator, the ecliptic and the Milky Way. It is the only constellation that crosses the ecliptic but is not part of the zodiac. The moon and planets can often be seen passing though the constellation's southern extension.

WHERE TO LOOK: HORIZON GRAPH

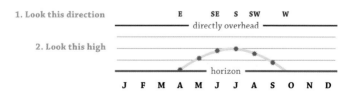

Shown 11pm DST/10pm ST on the 15th. If viewing earlier/later, adjust 1 month for every 2 hours. (At 9pm in Jun, use May; at 1am in Jun, use Jul.) See page 10 for more info.

Stars in Ophiucus

	LIGHT YEARS	MAGNITUDE	NAME ORIGIN
Ras Alhague (α)	47	2.1	Arabic for "head of the serpent collector"
Cebelrai (β)	82	2.8	Arabic for "the shepherd's dog"
RS Ophiuchi	5000	12.5*	Modern catalog designation

Every star is ranked in terms of brightness, or magnitude. The brighter the star, the lower the magnitude. Stars with negative magnitudes are very bright. The sixth magnitude is the limit of human vision; the brightest stars are first magnitude or less.

* Varies from 9.6 to 13.5, rarely flares to 5.0

Ras Alhague (α): The constellation's brightest star, with a magnitude of 2.1.

Cebelrai (β): Its name refers to a small, forgotten Arab constellation made up of several stars around Ophiucus's shoulder.

RS Ophiuchi: Normally invisible to the naked eye, this star is a member of a rare class of stars called recurrent novas. Recurrent novas are star systems that are usually very faint but erupt from time to time, increasing their brightness several hundredfold for several months before dimming again. RS Ophiuchi is one of only two recurrent novas that is ever visible to the naked eye (the other is T Coronae Borealis). It erupted in 1898, 1933, 1958, 1967, 1985 and 2006, each time reaching a magnitude of about 5.0—not at all bright, but still visible. Another eruption could happen at any time.

The Milky Way is visible in the southern and eastern extensions of the constellation, and the southern extension points roughly toward of the center of our galaxy.

Mythology/History:

Ophiucus has been recognized since antiquity as a man holding a large serpent (Serpens, page 127). In classical mythology, the figure has been associated with many different figures. Most commonly, however, the constellation is associated with Asclepius, the legendary healer. Asclepius was the mortal son of the gods Apollo and Coronus. He was raised by the centaur, Chiron, who taught him the healing arts. One day Asclepius killed a large snake with his staff. To his surprise, another snake approached and revived the first with herbs.

Asclepius began using these same herbs to treat the sick and became such a gifted healer that he could bring life back to the dead. Hades, god of the underworld, became worried that Asclepius's skill would mean an end to new souls entering his domain. He persuaded Zeus to strike down Asclepius with a thunderbolt and decree that humans must remain mortal. Enraged by the murder of his son, Apollo retaliated by killing the Cyclops who forged the thunderbolts for Zeus.

To appease Apollo, Zeus granted Asclepius immortality and placed him among the stars, holding the serpent. Asclepius is sometimes considered the god of medicine. The original, ancient Hippocratic oath began with the invocation "I swear by Apollo and by Asclepius." To this day, two serpents wound around a staff is a symbol of medicine.

Tang

Unukalhai

Leiolepis

WHERE TO LOOK: OVERHEAD MAP

Face south and hold book directly overhead

SPRING

SUMMER

FALL

WINTER

below the horizon

below the horizon

✻ = SERPENS   ⌐⌐ = BIG DIPPER   ✷ = NORTH STAR   ● = OVERHEAD   ⌐ = HORIZON

Maps depict mid-season at 11pm DST/10pm ST (May 5, Aug 5, Nov 5 and Feb 5).
See page 9 for other dates when maps are exact.

SERPENS *(SER-penz)*

English Name: the serpent

Size: medium, 23rd largest

When to Look: most prominent in June and July and visible in the late evening sky from May through September

Notes: This unique constellation is divided into two parts, the head and the tail of the snake, called Serpens Caput and Serpens Cauda, respectively. The two halves of the snake are separated by Ophiucus, the Serpent Holder. Though it stretches across a large area of the sky, it is a dim constellation that can be difficult to spot and trace. The head and the tail contain just one third-magnitude star each.

WHERE TO LOOK: HORIZON GRAPH

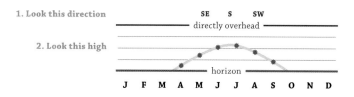

1. Look this direction

SE   S   SW

directly overhead

2. Look this high

horizon

J F M A M J J A S O N D

Shown 11pm DST/10pm ST on the 15th. If viewing earlier/later, adjust 1 month for every 2 hours. (At 9pm in Jun, use May; at 1am in Jun, use Jul.) See page 10 for more info.

Stars in Serpens

	LIGHT YEARS	MAGNITUDE	NAME ORIGIN
Unukalhai (α)	73	2.6	Arabic for "serpent's neck"
Tang (β)	62	3.2	Chinese reference to the Tang Imperial Dynasty
Leiolepis (μ)	156	3.5	Greek for "smooth scaled"

Every star is ranked in terms of brightness, or magnitude. The brighter the star, the lower the magnitude. Stars with negative magnitudes are very bright. The sixth magnitude is the limit of human vision; the brightest stars are first magnitude or less.

Unukalhai (α): The brightest star in the constellation, its name is Arabic for "serpent's neck." More rarely, it is called *Cor Serpenti*, which is Latin for "heart of the serpent."

Tang (β): The second-brightest star in the constellation and the brightest star in the tail. Its common name, which is not widely used, comes from Chinese and refers to the Tang Imperial Dynasty.

Leiolepis (μ): The southernmost star in *Serpens Caput*, its name is Greek for "smooth scaled."

The Milky Way passes through the far eastern edge of the constellation.

Mythology/History:

This constellation dates back to the ancient Greeks, and possibly even farther. It is a unique constellation in that it is divided into two parts, Serpens Caput (the head of the snake) and Serpens Cauda (the tail of the snake). Originally, these stars were considered to be part of the constellation Ophiucus, the Serpent Holder. Over the ages they have gradually come to be identified separately.

The practice of drawing distinct boundaries between constellations is a modern one, and was not formalized until the 1930s when the International Astronomical Union defined the 88 modern constellations. It was at this time that Serpens was clearly divided into two distinct regions separated by the constellation Ophiucus.

Despite this fragmentation, modern astronomers respect the tradition by considering these two areas of the sky to form a single constellation. Ophiucus is usually considered to be the legendary healer Asclepius, who was so gifted that he could bring the dead back to life. (For his story, page 125).

In ancient Greek culture, snakes were a symbol of rebirth. As they grow, snakes shed their skin, emerging soft and shiny, as if newborn. Because of this, snakes were an ancient Greek symbol of medicine, and they remain part of the emblem of the medical profession to this day.

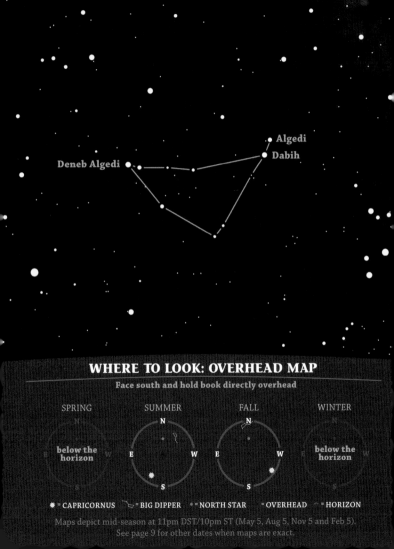

Algedi

Dabih

Deneb Algedi

WHERE TO LOOK: OVERHEAD MAP

Face south and hold book directly overhead

SPRING	SUMMER	FALL	WINTER
below the horizon	N E W S	N E W S	below the horizon

✻ = CAPRICORNUS   ⌐ = BIG DIPPER   • = NORTH STAR   ▫ = OVERHEAD   — = HORIZON

Maps depict mid-season at 11pm DST/10pm ST (May 5, Aug 5, Nov 5 and Feb 5).
See page 9 for other dates when maps are exact.

CAPRICORNUS *(CAP-rih-CORN-us)*

zodiac

English Name: the goat

Size: medium, 40th largest

When to Look: most prominent in August and September and visible in the late evening sky from July through October

Notes: A faint, southerly constellation with no truly bright stars. Its main stars form a somewhat lopsided triangle standing on its point. It is a bit tricky to spot, especially for northerners, and its odd but simple shape is fairly easy to trace. Among the constellations of the zodiac, only Cancer is harder to make out from our latitudes. As it is located along the ecliptic, the moon and planets can often be seen here.

WHERE TO LOOK: HORIZON GRAPH

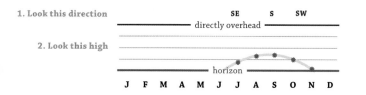

1. Look this direction

2. Look this high

Shown 11pm DST/10pm ST on the 15th. If viewing earlier/later, adjust 1 month for every 2 hours. (At 9pm in Aug, use Jul; at 1am in Aug, use Sep.) See page 10 for more info.

Stars in Capricornus

	LIGHT YEARS	MAGNITUDE	NAME ORIGIN
Deneb Algedi (δ)	39	2.9	Arabic for "tail of the goat"
Dabih (β)	300	3.4	Arabic for "lucky star of the sacrifice"
Algedi (α)	109 690	3.6 4.2	Arabic for "kid"

Every star is ranked in terms of brightness, or magnitude. The brighter the star, the lower the magnitude. Stars with negative magnitudes are very bright. The sixth magnitude is the limit of human vision; the brightest stars are first magnitude or less.

Deneb Algedi (δ): The brightest star in the constellation, its name is Arabic for "tail of the goat." The planet Neptune was discovered near this star in 1846 by German astronomer Johann Galle.

Dabih (β): The star's name probably comes from an Arabic phrase meaning "lucky star of the sacrifice," referring to the ritual sacrifice of a goat.

Algedi (α): An optical double. What appears at first glance to be a single star is actually two stars which, while very far apart from each other, lie in almost exactly the same direction from Earth. Together the two stars appear to shine with a magnitude of 3.1. Binoculars or very good eyesight can separate the magnitude 3.6 star from the magnitude 4.2 star. While nearly as bright, and appearing to be a close neighbor, the magnitude 4.2 is actually over 6 times as far away.

Ecliptic: The sun is found in Capricornus from January 18 to February 13. The tenth constellation in the zodiac, its astrological period runs from December 22 to January 19.

Mythology/History:

An ancient constellation, it was pictured on Babylonian tablets 3,000 years ago. In the past, the sun reached its most southerly point (the southern solstice) while in this constellation; at that time, the sun was directly over-head at 23.5 degrees south latitude. To this day, this line is known as the Tropic of Capricorn, even though the sun is now in Sagittarius on the solstice.

The word *Capricornus* is Latin for "horned goat," but the Babylonians depicted the constellation as a "sea goat," with the forequarters of a goat and the tail of a fish. In classical Greek mythology, the constellation is often associated with the god Pan. Pan is depicted as having a man's torso and the horns and legs of a goat.

According to Greek mythology, when the monster Typhon challenged the gods of Olympus, Pan sounded the alarm by blowing into a conch shell. Most of the gods fled from Typhon by changing themselves into animals. Pan evaded the monster by transforming his forequarters into those of a goat, then jumping into the river and transforming his hindquarters into a fish tail. After a tremendous struggle, Zeus defeated Typhon. Grateful for Pan's warning call, and amused by his clever escape, Zeus placed the sea goat in the stars. The astrological symbol for Capricornus is (♑) and represents the horns and fish tail of the sea goat.

γ **Sagittae**

Sham

WHERE TO LOOK: OVERHEAD MAP

Face south and hold book directly overhead

SPRING

SUMMER

FALL

WINTER

N

N

N

N

E

W

E

W

E

W

E

W

S

S

S

S

below the horizon

✳ = SAGITTA ⌐ʹ = BIG DIPPER ▪ = NORTH STAR ▫ = OVERHEAD ▬ = HORIZON

Maps depict mid-season at 11pm DST/10pm ST (May 5, Aug 5, Nov 5 and Feb 5).
See page 9 for other dates when maps are exact.

SAGITTA *(suh-JIT-uh)*

English Name: the arrow

Size: very small, 86th largest

When to Look: most prominent from July through September and visible in the late evening sky from June through November

Notes: Sagitta is one of the ancient constellations and although small and dim, its compact shape really does look like an arrow, making it easy to trace. Sitting in the Milky Way, toward the southern end of the Summer Triangle, it is also fairly easy to locate, but only when the sky is clear and dark. Because it is so small, it makes a good target for binoculars, as the whole constellation will fit in a single field of view.

WHERE TO LOOK: HORIZON GRAPH

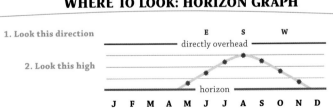

1. Look this direction

2. Look this high

Shown 11pm DST/10pm ST on the 15th. If viewing earlier/later, adjust 1 month for every 2 hours. (At 9pm in Jun, use May; at 1am in Jun, use Jul.) See page 10 for more info.

Stars in Sagitta

	LIGHT YEARS	MAGNITUDE	NAME ORIGIN
γ **Sagittae**	275	3.8	No common name
Sham (α)	475	4.4	Arabic for "arrow"

Every star is ranked in terms of brightness, or magnitude. The brighter the star, the lower the magnitude. Stars with negative magnitudes are very bright. The sixth magnitude is the limit of human vision; the brightest stars are first magnitude or less.

γ **Sagittae**: The brightest star in the constellation, with a magnitude of just 3.8, it marks the tip of the arrow.

Sham (α): The only star in the constellation that has a common name; *sham* is the Arabic word for "arrow."

The Milky Way passes through Sagitta.

Mythology/History:

Although tiny and dim, this constellation has been recognized as an arrow since antiquity. In fact, the constellation has been associated with nearly every famous arrow in Greek mythology. Various Greek authors identified these stars with Eros (Cupid), Hercules, the centaur Chiron and Apollo, with different stories about each. Nevertheless, two stories are commonly associated with these stars.

In one story, these stars represent an arrow shot by Hercules to rescue Prometheus. Prometheus was one of the Titans, ancient giants who ruled the Earth before the Olympian gods. In fact, Prometheus created man, molding him in the image of the gods. It was Prometheus who stole fire from Zeus and gave it to mankind. Angered by this theft, Zeus had Prometheus chained to a rock to be eternally punished. Each day an eagle came and fed on Prometheus's liver. As he was immortal, his liver regrew overnight, only to be consumed again. Hercules saved Prometheus by killing the eagle with an arrow.

In the other story, the constellation represents the arrow that Apollo used to avenge his son, Asclepius. Zeus struck Asclepius down with a thunderbolt because his power as a healer began to threaten the gods. In retaliation, Apollo shot and killed the Cyclops, a one-eyed giant that forged Zeus's thunderbolts.

Sualocin
Job's Coffin
Rotanev

WHERE TO LOOK: OVERHEAD MAP

Face south and hold book directly overhead

SPRING	SUMMER	FALL	WINTER
below the horizon			below the horizon

✳ = DELPHINUS   ⌐⌐ = BIG DIPPER   · = NORTH STAR   □ = OVERHEAD   ⌒ = HORIZON

Maps depict mid-season at 11pm DST/10pm ST (May 5, Aug 5, Nov 5 and Feb 5).
See page 9 for other dates when maps are exact.

DELPHINUS *(del-FIN-us)*

English Name: the dolphin

Size: very small, 69th largest

When to Look: most prominent in August and September and visible in the late evening sky from June through November

Notes: A charming little constellation. Small and dim, but easy to trace because its stars are so close together. It actually looks like a tiny dolphin arching its back as it leaps across the sky. It makes a very nice target for binoculars, as most of the constellation is visible in a single field of view.

WHERE TO LOOK: HORIZON GRAPH

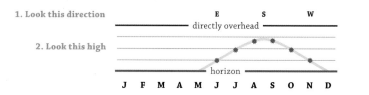

1. Look this direction

2. Look this high

E   S   W
directly overhead

horizon

J F M A M J J A S O N D

Shown 11pm DST/10pm ST on the 15th. If viewing earlier/later, adjust 1 month for every 2 hours. (At 9pm in Jul, use Jun; at 1am in Jul, use Aug.) See page 10 for more info.

Stars in Delphinus

	LIGHT YEARS	MAGNITUDE	NAME ORIGIN
Sualocin (α)	240	3.8	Reversed Latin for "Niccolo"
Rotanev (β)	97	3.5	Reversed Latin for "Cacciatore"

Every star is ranked in terms of brightness, or magnitude. The brighter the star, the lower the magnitude. Stars with negative magnitudes are very bright. The sixth magnitude is the limit of human vision; the brightest stars are first magnitude or less.

Sualocin & Rotanev (α & β): Peculiar names that appeared in an 1814 catalog from the Palermo Observatory in Italy. Read backwards, the names of the stars spell Nicolaus Venator, the Latinized form of the Italian name Niccolo Cacciatore, the assistant to the director of the Palermo Observatory, who was immortalized in the heavens by his grateful boss. Together with Cor Caroli in Canes Venatici (page 61), these stars are the only ones named after a real person in modern times.

Job's Coffin: A nickname for the four main stars that make up the body of the dolphin. Although it is now fairly well-known, no one knows where this unusual name came from or when it came into use.

The Milky Way passes through the northwest corner of the constellation.

Mythology/History:

The constellation is associated with two classical myths. In one myth, Poseidon decided that he wished to marry and he began to court the sea nymph Amphitrite. Out of modesty, and a desire to remain a maiden, Amphitrite fled to Mount Atlas. Poseidon sent out many messengers to seek her. It was the dolphin that found Amphitrite and convinced her to return and accept Poseidon's hand in marriage. In so doing, the dolphin became Poseidon's favorite subject, and he placed it among the stars.

The other myth associated with Delphinus tells of the Greek musician Arion of Lesbos. Arion traveled widely and earned a small fortune because of his musical talents. While sailing home with a large sum after winning a contest, the ship's crew conspired to murder him and steal his prize; Arion asked to play one final song. The crew granted this request, and his song was so beautiful that it attracted a pod of dolphins. Seeing this, Arion threw himself into the sea. Thinking him lost, the crew continued on their trip home, dividing the fortune between them. One of the dolphins, however, rescued Arion and carried him home to Lesbos faster than the ship could travel.

When the ship arrived, the crew was greeted by Arion and were severely punished. It is said that the dolphin was placed in the sky in recognition of this just outcome.

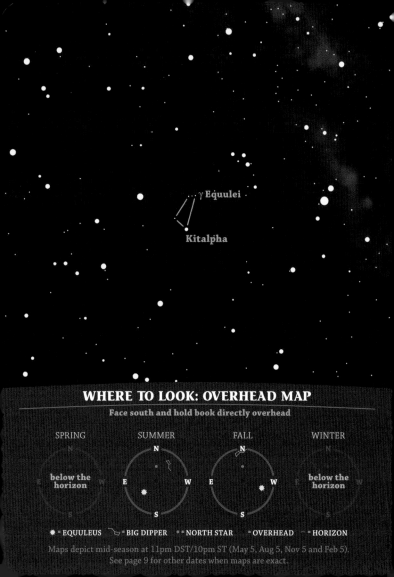

γ Equulei

Kitalpha

WHERE TO LOOK: OVERHEAD MAP

Face south and hold book directly overhead

SPRING | SUMMER | FALL | WINTER

below the horizon | | | **below the horizon**

✷ = EQUULEUS   〜 = BIG DIPPER   • = NORTH STAR   ▪ = OVERHEAD   ⌒ = HORIZON

Maps depict mid-season at 11pm DST/10pm ST (May 5, Aug 5, Nov 5 and Feb 5).
See page 9 for other dates when maps are exact.

EQUULEUS *(eh-QUOO-lee-us)*

English Name: the little horse

Size: very small, 87th largest

When to Look: most prominent in July and August and visible in the late evening sky from June through October

Notes: The second smallest of all the constellations and the smallest constellation visible from this region. Although tiny and dim, this constellation was known to the ancient Greeks. It is a faint but recognizable group of stars that actually does look a bit like an upside-down horse's head. Its brightest star, Kitalpha, marks the ear. Its compact shape and dim stars make it a good target for binoculars.

WHERE TO LOOK: HORIZON GRAPH

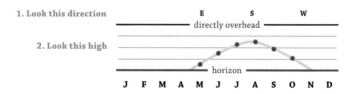

1. Look this direction

2. Look this high

Shown 11pm DST/10pm ST on the 15th. If viewing earlier/later, adjust 1 month for every 2 hours. (At 9pm in Jun, use May; at 1am in Jun, use Jul.) See page 10 for more info.

Stars in Equuleus

	LIGHT YEARS	MAGNITUDE	NAME ORIGIN
Kitalpha (α)	150	3.9	Arabic for "part of a horse"
γ **Equulei**	68	4.7	No common name

Every star is ranked in terms of brightness, or magnitude. The brighter the star, the lower the magnitude. Stars with negative magnitudes are very bright. The sixth magnitude is the limit of human vision; the brightest stars are first magnitude or less.

Tiny and dim, with just one fourth-magnitude star, this constellation is a good candidate for viewing in binoculars. Most binoculars have a wide enough field of view to take in the entire constellation at once.

Kitalpha (α): The brightest star in the constellation and the only one with a common name. Its name means "part of a horse" in Arabic.

γ **Equulei**: A faint star. Binoculars, or truly exceptional eyesight, reveal that it has a sixth-magnitude companion. The system lies 68 light years from Earth.

Mythology/History:

Although the constellation is ancient, there are no myths clearly associated with it. Ptolemy cataloged this constellation's four brightest stars in his treatise, the *Almagest,* in 150 AD, but did not otherwise discuss it. Its name, the "little horse," distinguishes it from nearby Pegasus. Some sources say that the constellation represents the horse Celeris, the brother of Pegasus, given to Castor (one of the twins of Gemini) by the god Hermes.

This appears to be a modern interpretation, however, as there are no references to this character in the original writings of the classical myths. Pegasus is said to be the son of Poseidon and the Gorgon Medusa, born when Medusa's blood dripped into the sea after she was beheaded by Perseus (page 167). There is no mention of another horse being born from either Medusa or Poseidon.

This constellation may be associated with Castor and Pollux because the Romans considered them the patrons of cavalry and horsemen. The constellation may have been named in honor of one of their horses, but specific mythology about these stars has been lost.

These stars are sometimes called *Equus Primus*, Latin for "the First Horse," because they rise in the east before the stars of Pegasus.

145

Anser

Coathanger Cluster (Cr 399)

WHERE TO LOOK: OVERHEAD MAP

Face south and hold book directly overhead

SPRING

SUMMER

FALL

WINTER

below the horizon

✴ = VULPECULA   ⌐⌐ = BIG DIPPER   • = NORTH STAR   ✴ = OVERHEAD   ⌐ = HORIZON

Maps depict mid-season at 11pm DST/10pm ST (May 5, Aug 5, Nov 5 and Feb 5).
See page 9 for other dates when maps are exact.

VULPECULA *(vul-PECK-yuh-luh)*

English Name: the little fox

Size: small, 55th largest

When to Look: most prominent from July through September and visible in the late evening sky from May through November

Notes: A faint, modern constellation located in the middle of the Summer Triangle—the large asterism defined by the bright stars of Deneb in Cygnus, Vega in Lyra, and Altair in Aquila. This position makes the constellation easy to locate, but its dim stars make it quite difficult to trace.

WHERE TO LOOK: HORIZON GRAPH

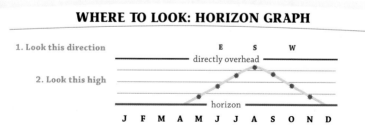

1. Look this direction

2. Look this high

E   S   W

directly overhead

horizon

J F M A M J J A S O N D

Shown 11pm DST/10pm ST on the 15th. If viewing earlier/later, adjust 1 month for every 2 hours. (At 9pm in Jun, use May; at 1am in Jun, use Jul.) See page 10 for more info.

Stars in Vulpecula

	LIGHT YEARS	MAGNITUDE	NAME ORIGIN
Anser (α)	296	4.4	Latin for "goose"

Every star is ranked in terms of brightness, or magnitude. The brighter the star, the lower the magnitude. Stars with negative magnitudes are very bright. The sixth magnitude is the limit of human vision; the brightest stars are first magnitude or less.

The constellation contains no bright stars and has little of interest to the naked-eye observer. Because it sits in the Milky Way, a few interesting features are visible in binoculars, especially the delightful Coathanger Star Cluster.

Anser (α): The brightest star in the constellation and the only one to have a common name or a Bayer (Greek letter) designation. Its name is Latin for "goose" and comes from Hevelius's original name for the constellation, *Vulpecula cum Ansere*, which means "little fox with a goose."

Coathanger Cluster (Cr 399): Visible to the naked eye as a faint fuzzy spot in the Milky Way. Binoculars reveal a tiny star cluster that looks like a miniature wooden coat hanger. The individual stars range from fifth- to seventh-magnitude. The combined magnitude of the cluster is 3.6.

Mythology/History:

A modern constellation created by Johannes Hevelius in the seventeenth century, Hevelius's original name for the constellation was *Vulpecula cum Ansere*, meaning "little fox with a goose," and depicted a fox with a goose in its jaws. Hevelius placed his new constellation in the sky just above Aquila, the eagle, and below Lyra, which used to be called the vulture and was still associated with that animal. He wrote that the location was fitting because, like the eagle and the vulture, the fox is cunning, voracious, fierce, rapacious and greedy. Three decades later, English astronomer John Flamsteed used the Latin name *Vulpecula et Anser* meaning "fox and goose," a common name for British pubs at the time. The constellation is now known simply as Vulpecula, but the goose survives as the name of the constellation's brightest star. Although *vulpecula* literally means "little fox" in Latin, it is often translated simply as "fox."

In 1967, Vulpecula became noteworthy as the site of the discovery of the first pulsar. A pulsar is a neutron star with a very strong magnetic field that spins very rapidly, giving off a strong pulse of radio waves every few seconds. The radio pulses from Vulpecula are a remarkably regular 1.337 seconds apart. Before the source of the pulses was discovered, the object was jokingly named Little Green Men 1.

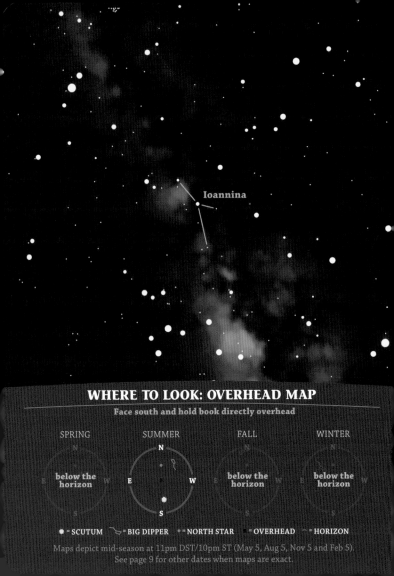

Ioannina

WHERE TO LOOK: OVERHEAD MAP

Face south and hold book directly overhead

SPRING

below the horizon

SUMMER

FALL

below the horizon

WINTER

below the horizon

✹ = SCUTUM ∾ = BIG DIPPER • = NORTH STAR ▪ = OVERHEAD ∾ = HORIZON

Maps depict mid-season at 11pm DST/10pm ST (May 5, Aug 5, Nov 5 and Feb 5).
See page 9 for other dates when maps are exact.

SCUTUM *(SCOOT-um)*

English Name: the shield

Size: very small, 84th largest

When to Look: most prominent in July and August and visible in the late evening sky from June through September

Notes: A small, dim, modern constellation. Although it contains only two stars of fourth magnitude, it lies in a rich region of the Milky Way and is a fine area to observe. Under dark skies, binoculars can reveal a number of faint star clusters and nebulae.

WHERE TO LOOK: HORIZON GRAPH

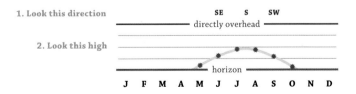

1. Look this direction

2. Look this high

Shown 11pm DST/10pm ST on the 15th. If viewing earlier/later, adjust 1 month for every 2 hours. (At 9pm in Jun, use May; at 1am in Jun, use Jul.) See page 10 for more info.

151

Stars in Scutum

	LIGHT YEARS	MAGNITUDE	NAME ORIGIN
Ioannina (α)	174	3.9	Greek for "John's"

Every star is ranked in terms of brightness, or magnitude. The brighter the star, the lower the magnitude. Stars with negative magnitudes are very bright. The sixth magnitude is the limit of human vision; the brightest stars are first magnitude or less.

Only two of the stars in Scutum are fourth magnitude, and only one bears a common name. While the few brighter stars here are of comparatively little interest, the area features rich star fields of the Milky Way and can be very interesting to view with binoculars.

Ioannina (α): The brightest star in the constellation, with a magnitude of 3.9. Its name is Greek for "John's," referring to King John III of Poland's coat of arms.

The Milky Way is visible in this region, as the constellation lies on the western edge of the Milky Way.

Mythology/History:

The constellation was created in the seventeenth century by Polish astronomer Johannes Hevelius in honor of his patron, King John III Sobieski of Poland. The original name for the constellation was *Scutum Sobiescianum,* meaning "Sobieski's Shield," and represents the monarch's coat of arms.

A brilliant military commander, Sobieski was the field commander of the Polish army. On November 11, 1673, he won an impressive battle against the Turkish army, capturing their fortress in Chocim. The battle was fought just one day after the death of King Michał I, and news of both events spread quickly through the country.

In seventeenth-century Poland, the monarchy was an elected position, and Sobieski presented himself as a candidate for the throne. A tremendously popular war hero, he ran nearly unopposed and was elected king in early 1674. Still facing threats from the Turks, Sobieski returned to his job as field commander and waged a 10-year campaign to secure Poland and much of the rest of Europe. In 1683, he signed the Treaty of Warsaw with Leopold I of the Holy Roman Empire. Later that year, outnumbered nearly two to one, he personally led a cavalry charge that broke the Turkish siege on Vienna and drove back the Turks. For this heroic victory, Sobieski was hailed by the Pope as "Savior of Vienna and Western European civilization," and immortalized in the stars.

Alphekka Meridiana

β Coronae Australis

WHERE TO LOOK: OVERHEAD MAP

Face south and hold book directly overhead

SPRING   SUMMER   FALL   WINTER

below the horizon   below the horizon   below the horizon

✳ = CORONA AUSTRALIS   ⌐ = BIG DIPPER   ⁎ = NORTH STAR   ▪ = OVERHEAD   ⌢ = HORIZON

Maps depict mid-season at 11pm DST/10pm ST (May 5, Aug 5, Nov 5 and Feb 5).
See page 9 for other dates when maps are exact.

CORONA AUSTRALIS *(cuh-ROE-nuh aw-STRAL-iss)*

English Name: the southern crown

Size: very small, 80th largest

When to Look: visible in the late evening sky in August

Notes: A very small, dim constellation in the southern skies. Although it was known to the ancient Greeks, it is very dim and difficult to spot. Its compact shape makes it fairly easy to trace if you can pick out its dim stars on the edge of the Milky Way.

WHERE TO LOOK: HORIZON GRAPH

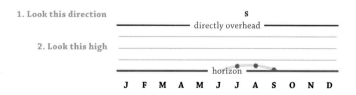

1. Look this direction

S

directly overhead

2. Look this high

horizon

J F M A M J J A S O N D

Shown 11pm DST/10pm ST on the 15th. If viewing earlier/later, adjust 1 month for every 2 hours. (At 9pm in Aug, use Jul; at 1am in Aug, use Sep.) See page 10 for more info.

Stars in Corona Australis

	LIGHT YEARS	MAGNITUDE	NAME ORIGIN
Alphekka Meridiana (α)	130	4.1	Arabic for "broken circle" and Latin for "midday" and "south"
β **Coronae Australis**	510	4.1	No common name

Every star is ranked in terms of brightness, or magnitude. The brighter the star, the lower the magnitude. Stars with negative magnitudes are very bright. The sixth magnitude is the limit of human vision; the brightest stars are first magnitude or less.

Alphekka Meridiana (α): The only star in the constellation that has a common name. It is tied with β Coronae Australis as the brightest star in the constellation—though both are fairly dim at magnitude 4.1. Its peculiar name is a mixture of Arabic and Latin. *Alphekka* is an Arabic name that is likely borrowed from the star in Corona Borealis. It derives from the Arabic word for "broken circle," which describes the shape of the crown. *Meridiana* is Latin for "midday" and is also used to mean "south," since the sun is due south at midday. Thus, Alphekka Meridiana is a bizarre rendering of the constellation's name.

β **Coronae Australis:** Appearing next to Alphekka Meridiana in the sky and matching it in brightness at magnitude 4.1, this star is four times as far away.

Mythology/History:

The constellation's name means "Southern Crown," but in classical depictions the constellation is depicted as a wreath. Wreathes of laurel and olive branches were important symbols of status and achievement in ancient Greece. Olympic champions were awarded crowns of olive branches to wear; these wreathes symbolized the pinnacle of athletic achievement and were considered the mark of a hero. An Olympic victory, symbolized by the olive crown, brought glory and recognition to both the athlete and his city.

Scholarly and literary achievement were recognized by the ancient Greeks with wreathes of laurel. When a student completed his scholarly training, he was awarded a crown of laurel leaves and berries called a "*bacca-laureate*," literally meaning "laurel berries." Poetic achievement was similarly recognized. To this day, people who achieve recognition in poetry or in scholarly pursuits are called "laureates."

In ancient Greece, baccalaurei usually did not marry, so that the demands of being a husband and a father would not interfere with their scholarly or literary pursuits. This tradition gave rise to the term bachelor, which over the ages has been extended to unmarried men in general. The constellation sits at the feet of Sagittarius (page 111), and some suggest that this is his crown, recognizing his skill as a musician and an athlete.

ε Microscopii   γ Microscopii

WHERE TO LOOK: OVERHEAD MAP
Face south and hold book directly overhead

SPRING   SUMMER   FALL   WINTER

below the horizon

N

E   W

S

below the horizon   **below the horizon**

✴ = MICROSCOPIUM   ⌐ = BIG DIPPER   ✱ = NORTH STAR   · = OVERHEAD   – = HORIZON

Maps depict mid-season at 11pm DST/10pm ST (May 5, Aug 5, Nov 5 and Feb 5).
See page 9 for other dates when maps are exact.

MICROSCOPIUM (my-cruh-SCOPE-ee-um)

English Name: the microscope

Size: small, 66th largest

When to Look: visible in the late evening sky from August through September

Notes: A small, dim, modern constellation. It is very difficult to spot or to trace. Its brightest stars are just of the fifth magnitude, putting it near the limit of naked-eye visibility. Under ideal conditions, it is fully visible throughout the United States and sits just below Capricornus, but its faint stars may be impossible to pick out so close to the horizon.

WHERE TO LOOK: HORIZON GRAPH

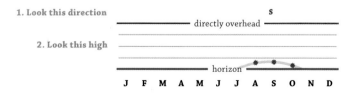

1. Look this direction

directly overhead

S

2. Look this high

horizon

J F M A M J J A S O N D

Shown 11pm DST/10pm ST on the 15th. If viewing earlier/later, adjust 1 month for every 2 hours. (At 9pm in Sep, use Aug; at 1am in Sep, use Oct.) See page 10 for more info.

Stars in Microscopium

	LIGHT YEARS	MAGNITUDE	NAME ORIGIN
γ **Microscopii**	220	4.7	No common name
ε **Microscopii**	165	4.7	No common name

Every star is ranked in terms of brightness, or magnitude. The brighter the star, the lower the magnitude. Stars with negative magnitudes are very bright. The sixth magnitude is the limit of human vision; the brightest stars are first magnitude or less.

None of the stars in Microscopium have common names and even the brightest are near the limit of naked-eye visibility. The constellation also contains little of interest for viewers with binoculars.

γ **Microscopii**: Based on its apparent motion, scientists have estimated that 3.8 million years ago this star would have been our closest neighbor, just 6 light years distant. At this distance, the star would have had an apparent magnitude of about −3, outshining everything in the sky except for the sun, moon and Venus.

ε **Microscopii**: The constellation's second-brightest star, it is just a tiny bit dimmer than its companion and also has a magnitude of 4.7.

Mythology/History:

This constellation was created by French astronomer Nicolas Louis de Lacaille in the eighteenth century in commemoration of the microscope. Like most of the 14 constellations that de Lacaille defined, it is very faint, named after an important instrument of science and looks nothing like its namesake. The Greeks considered these stars to be one part of a vast area known as "the sea," not part of a specific constellation.

The constellation was named in honor of the compound microscope, which uses multiple lenses to magnify an image. Magnifying lenses have their origins in "reading stones," which originated in Islamic Spain about 1,000 years ago. More sophisticated optics were first produced by European Renaissance inventors in the sixteenth century and helped usher in the scientific revolution.

Galileo Galilei was one of the first scientists to use a compound microscope and is sometimes credited as its inventor. Galileo was a member of the newly formed *Accademia dei Lincei,* one of the most famous scientific academies. The name means "academy of the lynx-eyed," a reference to the keen powers of observation required by science. The Linceans named Galileo's device the "microscope," which stems from the Greek words *micron* (small) and *skopein* (to look at). The name is similar to the "telescope," also named by the Linceans; de Lacaille honored it in a constellation too.

161

THE FALL SKY
(overhead, facing south)

This map shows the fall sky as it appears at 1am on Oct 5, 10pm on Nov 5, and 8pm on Dec 5. For other times this chart can be used, see page 9.

AS WE MOVE INTO AUTUMN, nightfall comes earlier each passing day. With many evenings still comfortably warm, the early nights invite us to spend the hours after darkness gazing into the cosmos. The fall sky is dominated by the constellations of the Andromeda Group—the half dozen constellations that represent the characters of the Greek legend of the Princess Andromeda (page 175). Look for the Great Square of Pegasus (page 165), and the "W" shape of Cassiopeia (page 169).

Fall also gives us one of our best views of the distant universe. On a clear night, you can spot the Andromeda Galaxy (page 173)—the most distant object visible with the naked eye.

Back in our own galaxy, look toward the constellation Perseus (page 177) for the so-called "Demon Star," Algol, and for the Algenib Star Cluster, which is wonderful in binoculars.

The fall constellations are in blue on the seasonal sky map to the left.

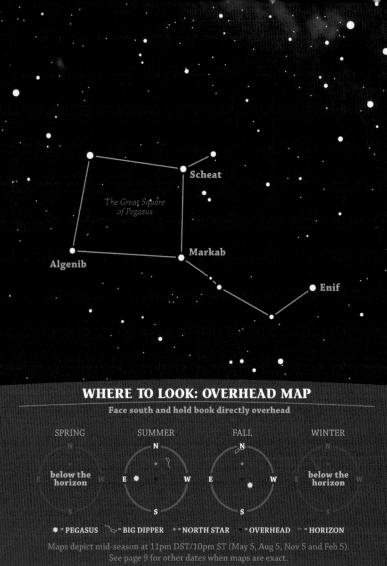

Scheat

The Great Square of Pegasus

Algenib

Markab

Enif

WHERE TO LOOK: OVERHEAD MAP

Face south and hold book directly overhead

SPRING

below the horizon

SUMMER

N

E   W

S

FALL

N

E   W

S

WINTER

below the horizon

✳ = PEGASUS   ⌐⌐ = BIG DIPPER   • = NORTH STAR   • = OVERHEAD   = HORIZON

Maps depict mid-season at 11pm DST/10pm ST (May 5, Aug 5, Nov 5 and Feb 5).
See page 9 for other dates when maps are exact.

PEGASUS *(PEG-us-us)*

English Name: the winged horse

Size: very large, 7th largest

When to Look: most prominent from September through November and visible in the late evening sky from July through December

Notes: The Great Square of Pegasus dominates the fall sky. Though not exceptionally bright, the four stars that mark the corners of the Great Square surround a relatively empty area of the sky, making it is easy to spot and trace. The star marking the northeast corner of the great square is officially part of the constellation Andromeda, but sits on the border between the two constellations and is usually shown in drawings of Pegasus.

WHERE TO LOOK: HORIZON GRAPH

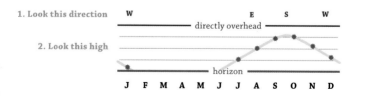

Shown 11pm DST/10pm ST on the 15th. If viewing earlier/later, adjust 1 month for every 2 hours. (At 9pm in Sep, use Aug; at 1am in Sep, use Oct.) See page 10 for more info.

Stars in Pegasus

	LIGHT YEARS	MAGNITUDE	NAME ORIGIN
Enif (ε)	700	2.4	Arabic for "nose"
Markab (α)	140	2.4	Arabic for "shoulder of the horse"
Scheat (β)	200	2.4	Arabic for "shin"
Algenib (γ)	335	2.8	Arabic for "the side"

Every star is ranked in terms of brightness, or magnitude. The brighter the star, the lower the magnitude. Stars with negative magnitudes are very bright. The sixth magnitude is the limit of human vision; the brightest stars are first magnitude or less.

Enif (ε): The brightest star in the constellation by a small margin, its name is Arabic for "nose" and marks the muzzle of the horse.

Markab (α): Marks the southwest corner of the Great Square.

Scheat (β): Marks the northwest corner of the Great Square. It is a red giant near the end of its life and has an unusually low surface temperature for a bright star (6,200° F) making it shine a deep red color, which is visible through binoculars.

Algenib (γ): Marks the southeast corner of the Great Square. At a distance of 335 light years it is the dimmest star in the Great Square.

The Great Square of Pegasus: A large square marked by the stars Markab, Scheat, Algenib and Alpheratz (in the constellation Andromeda, page 173), enclosing a relatively dark region of the sky. The Great Square is the most prominent feature in the fall sky.

Mythology/History:

An ancient constellation that may trace its history back to images of winged horses on early tablets and vases found in the Euphrates river valley. The earliest Greek descriptions of Pegasus and of the constellation make no mention of his wings. By the time of Ptolemy, the wings were the defining feature of the great horse.

In Greek mythology, Pegasus was the son of Poseidon and the mortal Medusa. In her youth, Medusa was famed for her beauty and especially renowned for her hair. As a young woman, she was seduced in the temple of Athena by Poseidon, the god of the sea and the god of horses. As punishment for her lack of chastity in the temple, Athena condemned Medusa to a life of grue-some ugliness. From that point on, her face was covered with dragon scales, her hands were made of brass, and she grew snakes for hair. Her appearance was so hideous that any mortal who looked at her was turned to stone. The hero Perseus slew Medusa while she slept by cutting off her head. As her head rolled to the ground, Pegasus sprang from her blood. After his unusual birth, Pegasus flew up to the heavens where he became the thundering steed who carried Zeus's lightning bolts. The constel-lation of Pegasus represents only the front quarters of the horse, seen upside down. One modern mythologist suggested that Pegasus was placed in the stars whole, but that his hindquarters later fell to Earth where they became the patron of politicians.

Cih

Caph

Schedar

WHERE TO LOOK: OVERHEAD MAP

Face south and hold book directly overhead

SPRING   SUMMER   FALL   WINTER

✴ = CASSIOPEIA   〜 = BIG DIPPER   •• = NORTH STAR   ▪ = OVERHEAD   ▪ = HORIZON

Maps depict mid-season at 11pm DST/10pm ST (May 5, Aug 5, Nov 5 and Feb 5).
See page 9 for other dates when maps are exact.

CASSIOPEIA *(CASS-ee-uh-PEE-uh)*

English Name: the Queen of Joppa

Size: medium, 25th largest

When to Look: most prominent from September through December and visible in the late evening sky throughout the year

Notes: A bright, distinctive constellation, it looks nothing like a woman. Rather, the constellation's main stars form a "W" or "M" in the sky. With a bit of imagination, this does look something like a chair, and the constellation is usually said to depict Cassiopeia sitting on her throne. Circumpolar above 40 degrees north latitude, it is visible for at least part of every night of the year from most of the United States.

WHERE TO LOOK: HORIZON GRAPH

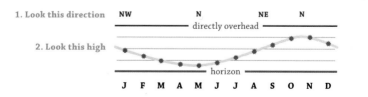

1. Look this direction

NW   N   NE   N

—— directly overhead ——

2. Look this high

—— horizon ——

J F M A M J J A S O N D

Shown 11pm DST/10pm ST on the 15th. If viewing earlier/later, adjust 1 month for every 2 hours. (At 9pm in Sep, use Aug; at 1am in Sep, use Oct.) See page 10 for more info.

Stars in Cassiopeia

	LIGHT YEARS	MAGNITUDE	NAME ORIGIN
Schedar (α)	230	2.2	Arabic for "beast"
Caph (β)	54	2.3	Arabic for "hand"
Cih (γ)	780	2.8*	Chinese for "whip"

Every star is ranked in terms of brightness, or magnitude. The brighter the star, the lower the magnitude. Stars with negative magnitudes are very bright. The sixth magnitude is the limit of human vision; the brightest stars are first magnitude or less.

* Varies from 2.15 to 3.4

Schedar (α): The brightest non-variable star in the constellation.

Caph (β): Its name means "hand" and is the Arabic name for the entire constellation.

Cih (γ): A member of a class of variable stars that spin very rapidly, throwing off rings of super-hot gases, which cause the star's magnitude to vary unpredictably. At its brightest, it outshines all of the other stars in the constellation. In 2008, the star shone with a magnitude of 2.15. In the 1940s, however, the star dimmed to a magnitude of just 3.4. Despite its prominence, the star has no Greek, Latin, or Arabic common name, so it is often referred to by its Chinese common name. The star was used as a navigational star during early manned space flights. Astronaut Gus Ivan Grissom gave the star the nickname "Navi," which is his own middle name spelled backwards.

The Milky Way passes though Cassiopeia.

Mythology/History:

Despite its bright stars and distinctive shape, Cassiopeia was a relatively late addition to the ancient constellations and has only one legend associated with it.

Cassiopeia represents the mythical Queen of Joppa, famous for her vanity. Wife of King Cepheus and the mother of Andromeda, Cassiopeia once claimed that she and her daughter were more beautiful than any of the sea nymphs that accompanied the sea god Poseidon's court. The sea nymphs complained to Poseidon, who punished Cassiopeia by sending a sea monster, Cetus, to ravage the coast of Joppa.

Cassiopeia and Cepheus consulted the Oracle, who told them that they had to sacrifice their beloved daughter to the monster to save their kingdom. Reluctantly, Cassiopeia and Cepheus had Andromeda chained to a rock on the coast in sacrifice. She was saved from Cetus by the hero, Perseus, who slew the monster and then took Andromeda's hand in marriage.

Poseidon placed Cassiopeia in the sky—but in a most undignified pose. Cassiopeia is said to be seated in her throne, but she rotates headfirst about the pole star—upside down half of the time—in eternal punishment for her vanity. All of the other main characters in the legend are also immortalized in constellations.

171

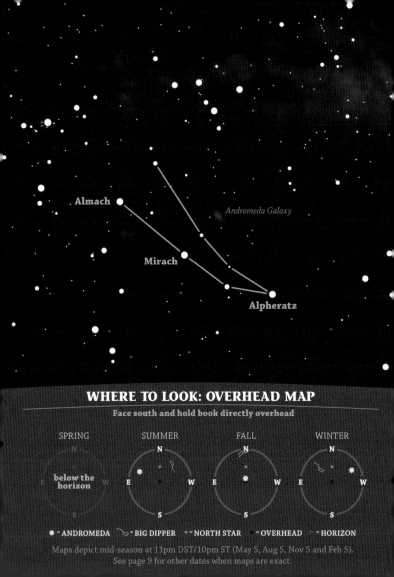

Almach

Mirach

Andromeda Galaxy

Alpheratz

WHERE TO LOOK: OVERHEAD MAP

Face south and hold book directly overhead

SPRING

below the horizon

SUMMER

N

E   W

S

FALL

N

E   W

S

WINTER

N

E   W

S

✳ = ANDROMEDA   ⌒ = BIG DIPPER   • = NORTH STAR   ▪ = OVERHEAD   = HORIZON

Maps depict mid-season at 11pm DST/10pm ST (May 5, Aug 5, Nov 5 and Feb 5).
See page 9 for other dates when maps are exact.

ANDROMEDA *(an-DRAH-mih-duh)*

English Name: the Princess of Joppa

Size: large, 19th largest

When to Look: most prominent from September through December and visible in the late evening sky from July through February

Notes: The main stars of the constellation form a long, curving "A" lying on its side, making it easy to recognize in the sky. It's more challenging to trace the entire constellation, which depicts the Princess Andromeda chained to a rock. The constellation is home to the most distant object visible to the naked eye, the Andromeda Galaxy, which is 2.5 million light years from Earth.

WHERE TO LOOK: HORIZON GRAPH

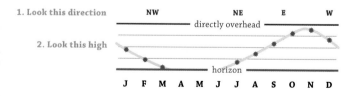

1. Look this direction

2. Look this high

Shown 11pm DST/10pm ST on the 15th. If viewing earlier/later, adjust 1 month for every 2 hours. (At 9pm in Sep, use Aug; at 1am in Sep, use Oct.) See page 10 for more info.

Stars in Andromeda

	LIGHT YEARS	MAGNITUDE	NAME ORIGIN
Alpheratz (α)	97	2.1	Arabic for "horse's shoulder"
Mirach (β)	200	2.1	Arabic for "girdle"
Almach (γ)	120	2.2	Refers to a caracal, a type of wildcat

Every star is ranked in terms of brightness, or magnitude. The brighter the star, the lower the magnitude. Stars with negative magnitudes are very bright. The sixth magnitude is the limit of human vision; the brightest stars are first magnitude or less.

Alpheratz (α): The brightest star in the constellation, it marks the northeast corner of the Great Square of Pegasus. Its Arabic name, which means "horse's shoulder," shows that this star was once considered to be part of the constellation Pegasus.

Mirach (β): The middle of the southern line of stars. Its name means "girdle," making it one of the few names that relates to the maiden Andromeda.

Almach (γ): The easternmost star in the southern line. The name apparently refers to the caracal, a wildcat native to much of Africa and the Middle East. It is a quadruple star system, but a telescope is needed to resolve its parts.

Andromeda Galaxy (M31): The most distant object in the universe visible to the naked eye, it is 2.5 million light years from Earth. Under dark skies, it appears as a faint, fuzzy oval northeast of Mirach. Thought to be a nebula prior to the twentieth century, it is now known to be a spiral galaxy similar to the Milky Way. It contains an estimated 200 billion stars and is 120,000 light years across. It has an apparent magnitude of about 3.5, but because it is so spread out, it can be more difficult to see than a magnitude 3.5 star.

Mythology/History:

In Greek mythology, Andromeda was the Princess of Joppa and the beautiful and beloved daughter of King Cepheus and Queen Cassiopeia. These three characters, and the constellations that bear their names, date back to the plays of Sophocles in the fifth century BC.

According to legend, Cassiopeia was very beautiful, but also boastful. She once claimed that she and her daughter were more beautiful than any of the sea nymphs who attended Poseidon's court. The nymphs complained to Poseidon, and he sent a sea monster, Cetus, to ravage the coast of Joppa. Andromeda was chained to a rock on the coast as a sacrifice to atone for her mother's boasting and save the kingdom.

The hero Perseus came upon the tragic scene and offered to save Andromeda, in exchange for her hand in marriage. Unhappy with the deal, but having no other option, her parents accepted. Perseus slew the monster and a wedding was performed on the spot.

Later, the goddess Athena placed Andromeda in the heavens to accompany her parents, who were placed there by Poseidon. Perseus and the monster Cetus are also immortalized in nearby constellations.

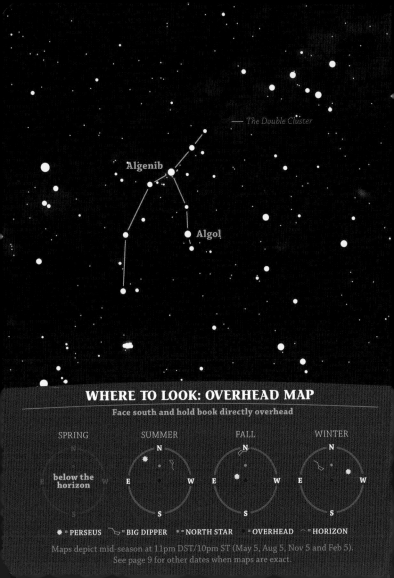

— *The Double Cluster*

Algenib

Algol

WHERE TO LOOK: OVERHEAD MAP

Face south and hold book directly overhead

SPRING	SUMMER	FALL	WINTER
below the horizon			

★ = PERSEUS   ∿ = BIG DIPPER   • = NORTH STAR   ▪ = OVERHEAD   ⌒ = HORIZON

Maps depict mid-season at 11pm DST/10pm ST (May 5, Aug 5, Nov 5 and Feb 5).
See page 9 for other dates when maps are exact.

PERSEUS *(PER-see-us)*

English Name: the hero

Size: medium, 24th largest

When to Look: most prominent from October through January and visible in the late evening sky from August through March

Notes: A fairly bright constellation situated across the Milky Way in the northern sky. Fairly easy to spot, but somewhat indistinct in shape. It contains a number of interesting objects visible with the naked eye, including the famous variable star Algol.

WHERE TO LOOK: HORIZON GRAPH

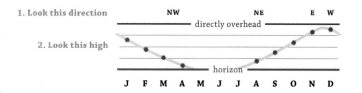

1. Look this direction

2. Look this high

Shown 11pm DST/10pm ST on the 15th. If viewing earlier/later, adjust 1 month for every 2 hours. (At 9pm in Sep, use Aug; at 1am in Sep, use Oct.) See page 10 for more info.

Stars in Perseus

	LIGHT YEARS	MAGNITUDE	NAME ORIGIN
Algenib (α)	590	1.8	Arabic for "the side"
Algol (β)	93	2.7*	Arabic for "head of the demon"

Every star is ranked in terms of brightness, or magnitude. The brighter the star, the lower the magnitude. Stars with negative magnitudes are very bright. The sixth magnitude is the limit of human vision; the brightest stars are first magnitude or less.

* Varies from 2.1 to 3.3 during eclipse

Algenib (α): The brightest star in the constellation, it is surrounded by a compact cluster of stars known either as Molette 20 or the Algenib Star Cluster. It is a fine target for binoculars.

Algol (β): One of the most famous variable stars in the sky, it is the prototype for a class of eclipsing binary stars called Algol-class variable stars. Algol variables are systems which include a large, bright primary star that is frequently eclipsed by a dimmer companion star. Algol varies in brightness every 3 days, so anyone can compare its brightness to other stars to see how much it has changed. In classical depictions, Algol marks the eye of the head of Medusa held by Perseus. For this reason it is sometimes called the Demon Star.

The Double Cluster (NGC 869 and NGC 884): First recorded by the ancient Greek astronomer Hipparchus, these two star clusters are just visible to the naked eye. Binoculars reveal a pair of rich star clusters each covering an area about the size of the full moon. Backed by the rich star fields of the Milky Way, the Double Cluster is a favorite target for amateur astronomers. The two clusters are separated by a few hundred light years, are about 65 light years across and are estimated to be between 12–20 million years old. NGC 884 is the brighter of the two and the more distant, lying 7,600 light years from Earth. NGC 869 is slightly dimmer, has fewer stars and lies 6,800 light years away.

Mythology/History:

This constellation is named for one of the most famous heroes in Greek mythology. The son of Zeus and the mortal Danaë, Perseus is best known for slaying the Gorgon Medusa and rescuing Princess Andromeda from the sea monster Cetus. Perseus is usually depicted as carrying the head of Medusa in his left hand.

According to legend, Perseus began his life of adventure to defend his mother, Danaë, from the unwanted advances of King Polydectes. The king offered to find a different bride if Perseus brought him the head of the Gorgon Medusa. Medusa was the one mortal of the three Gorgons—sisters so ugly that any mortal who looked at them would be turned to stone. Polydectes assumed that Perseus would die in the attempt, allowing him to court Danaë.

Nevertheless, Perseus succeeded, with the help of the gods. Flying on the winged sandals of Hermes, Perseus used the shield of Athena as a mirror to avert his gaze as he crept up on the sleeping Medusa and cut off her head. The winged horse Pegasus arose when Medusa's blood dripped into the ocean. Perseus mounted the steed and fled Medusa's immortal sisters. On his way, Perseus came upon Andromeda and rescued her, later winning her hand in marriage. The other characters in the story of Andromeda are also immortalized in nearby constellations.

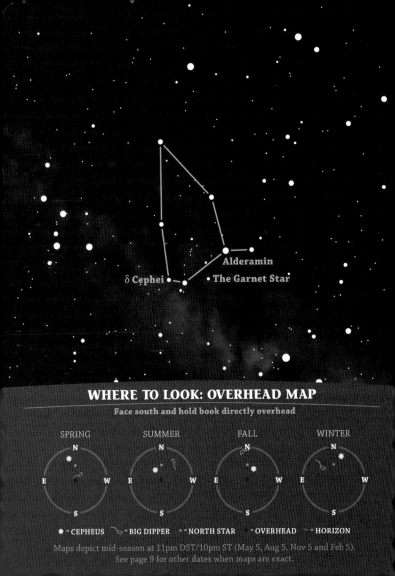

δ **Cephei**
Alderamin
The Garnet Star

WHERE TO LOOK: OVERHEAD MAP

Face south and hold book directly overhead

SPRING   SUMMER   FALL   WINTER

☀ = CEPHEUS   ⌐ = BIG DIPPER   • = NORTH STAR   ☀ = OVERHEAD   ⌒ = HORIZON

Maps depict mid-season at 11pm DST/10pm ST (May 5, Aug 5, Nov 5 and Feb 5).
See page 9 for other dates when maps are exact.

CEPHEUS *(SEE-fee-us)*

English Name: The King of Joppa

Size: medium, 27th largest

When to Look: most prominent from July through November and visible in the late evening sky throughout the year

Notes: A far-northern constellation that can be seen throughout the year, circling around the pole. Though it contains no particularly bright stars, its distinctive shape, which looks like a child's drawing of a house, makes it easy to identify.

WHERE TO LOOK: HORIZON GRAPH

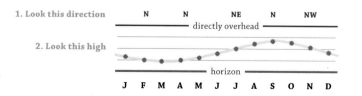

1. Look this direction

N   N   NE   N   NW

— directly overhead —

2. Look this high

— horizon —

J F M A M J J A S O N D

Shown 11pm DST/10pm ST on the 15th. If viewing earlier/later, adjust 1 month for every 2 hours. (At 9pm in Sep, use Aug; at 1am in Sep, use Oct.) See page 10 for more info.

Stars in Cepheus

	LIGHT YEARS	MAGNITUDE	NAME ORIGIN
Alderamin (α)	49	2.4	Arabic for "the right arm"
δ **Cephei**	891	4.0*	No common name
The Garnet Star (μ)	1,550	4.4**	Named for its striking color

Every star is ranked in terms of brightness, or magnitude. The brighter the star, the lower the magnitude. Stars with negative magnitudes are very bright. The sixth magnitude is the limit of human vision; the brightest stars are first magnitude or less.

* Varies from 3.5 to 4.5   ** Varies from 3.6 to 5.1

In about 1,000 years, Errai (γ Cephei) will be closer to the celestial pole than Polaris. After that, Alvahet (ι Cephei) will be the pole star. In 7,500 years Alderamin (α-Cephei) will follow suit, with Polaris about 26 degrees away.

Alderamin (α): The brightest star in the constellation. It is beginning to evolve into a red giant as it uses up its store of hydrogen.

δ **Cephei**: Delta Cephei is the prototype for an important class of stars, Cepheid variables, which are used as cosmic yardsticks by astronomers. Cepheid variables are supergiant stars that change in brightness over a fairly short period of time. The light output of these stars is directly correlated to their period of variation. Brighter Cepheid variables have a longer cycle between maximum and minimum brightness. Astronomers use the period of variation to precisely determine a Cepheid variable's light output. Because a star's apparent magnitude (how bright it appears to us on Earth) is a function of its light output and its distance from Earth, astronomers can use this information to calculate the distances to these stars very accurately.

The Garnet Star (μ): One of the most intensely colored stars known, its brilliant orange-red color is visible in binoculars. Interestingly, this star lies on the north celestial pole axis of Mars, making it the Red Planet's North Star. It is a variable star.

Mythology/History:

Cepheus is a member of the Andromeda group of constellations, which includes all of the main characters in the Andromeda legend. In fact, the tale of Andromeda is the only classical myth to be so fully represented in the constellations, suggesting that the stars themselves may have been the inspiration for the story and its characters.

According to the legend, these stars represent the king of the ancient land of Joppa, which was located on the northern coast of Africa. Cepheus was the husband of the vain Queen Cassiopeia and father of the beautiful Princess Andromeda. Cassiopeia boasted that she and her daughter were more beautiful than the Nereids, the sea nymphs who attended the court of Poseidon, the god of the sea. Poseidon responded by sending the sea monster Cetus to destroy the kingdom as a punishment for Cassiopeia's vain boasting.

Powerless in the face of divine wrath, Cepheus reluctantly had Andromeda, his only child, chained to a rock in sacrifice. She was saved by the hero, Perseus, in exchange for her hand in marriage. The couple gave Cepheus a grandson and heir before departing for Argos.

It is said that Poseidon placed Cepheus in the heavens not to honor any particular act of his own, but to complement the other actors in the tale.

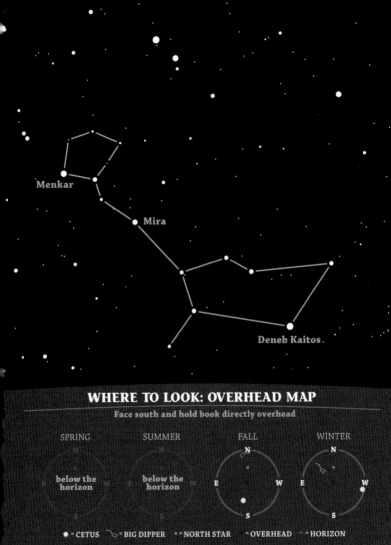

Menkar

Mira

Deneb Kaitos

WHERE TO LOOK: OVERHEAD MAP

Face south and hold book directly overhead

SPRING	SUMMER	FALL	WINTER
below the horizon	below the horizon		

✳ = CETUS ⟍ = BIG DIPPER ✶ = NORTH STAR ▬ = OVERHEAD ⌒ = HORIZON

Maps depict mid-season at 11pm DST/10pm ST (May 5, Aug 5, Nov 5 and Feb 5).
See page 9 for other dates when maps are exact.

CETUS *(SEE-tus)*

English Name: the sea monster

Size: very large, 4th largest

When to Look: most prominent in November and visible in the late evening sky from September through January

Notes: A huge constellation straddling the celestial equator. It contains a few brighter stars which are fairly easy to spot, but is very spread out making it somewhat tricky to trace. It also contains Mira, the first variable star ever recorded.

WHERE TO LOOK: HORIZON GRAPH

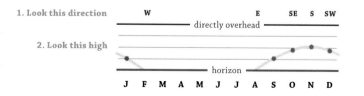

1. Look this direction

2. Look this high

Shown 11pm DST/10pm ST on the 15th. If viewing earlier/later, adjust 1 month for every 2 hours. (At 9pm in Oct, use Sep; at 1am in Oct, use Nov.) See page 10 for more info.

Stars in Cetus

	LIGHT YEARS	MAGNITUDE	NAME ORIGIN
Deneb Kaitos (β)	96	2.0	Arabic for "tail of the sea monster"
Menkar (α)	220	2.5	Arabic for "nostrils"
Mira (ο)	420	6.0*	Latin for "wonderful"

Every star is ranked in terms of brightness, or magnitude. The brighter the star, the lower the magnitude. Stars with negative magnitudes are very bright. The sixth magnitude is the limit of human vision; the brightest stars are first magnitude or less.

* Varies from 2.0 to 10.0

Deneb Kaitos (β): The brightest star in the constellation, it is occasionally called *Diphda*, an Arabic word meaning "second frog." The Arabs considered the star Fomalhaut, in Piscis Austrinus, the first frog.

Menkar (α): The constellation's second-brightest (non-variable) star, on the opposite side of the constellation from Deneb Kaitos.

Mira (ο): One of the most famous variable stars in the sky and the first variable star ever to be recorded, it is the prototype for the Mira class of variable stars. Mira variables are red giant stars that swell and contract over a period of months to years, varying their light output by as much as 10 magnitudes. Over a period of 332 days, Mira can vary its light output over 1,000 fold. At its brightest, Mira can reach magnitude 2.0, making it quite prominent in the sky, but it drops to magnitude 10.0 at its dimmest, requiring a telescope to see. The star plays an important role in the history of astronomy. Its discovery in 1596 by German astronomer David Fabricius gave a serious challenge to the Ptolemaic view of the universe, which held that the heavens were fixed and unchanging. The star's name comes from the Latin word for "wonderful," from which we get the word "miracle."

Mythology/History:

Cetus is a character in the Andromeda legend; unlike the rest of the constellations in the group, Cetus sits far to the south, straddling the celestial equator. It is also the oldest constellation in the group, and may have originated in the Egyptian zodiac, where it represented the crocodile.

In Greek mythology, the constellation represents the sea monster sent by the sea god Poseidon to ravage the coast of Joppa. Queen Cassiopeia bragged that she and her daughter were more beautiful than Poseidon's daughters, the Nereids. To punish her for her vain boasting, Poseidon sent Cetus to ravage the kingdom. Cassiopeia and her husband, Cepheus, consulted the Oracle and were told that to save their kingdom they must sacrifice their beloved daughter, Andromeda, to the monster. Resigned to their fate, they had Andromeda chained to a rock on the coast. As Cetus approached to claim his prize, the hero Perseus arrived and slew the monster, winning Andromeda's hand in marriage.

Sea monsters are unusual in Greek mythology, and there are few descriptions of the beast in classical literature. The word *cetus* actually means "whale" in Latin, but this creature was clearly no gentle sea mammal. Artistic renderings of Cetus usually show him as a fearsome sea dragon.

Fomalhaut

WHERE TO LOOK: OVERHEAD MAP

Face south and hold book directly overhead

SPRING	SUMMER	FALL	WINTER
below the horizon			below the horizon

✴ = PISCIS AUSTRINUS ☇ = BIG DIPPER • = NORTH STAR • = OVERHEAD ⌒ = HORIZON

Maps depict mid-season at 11pm DST/10pm ST (May 5, Aug 5, Nov 5 and Feb 5).
See page 9 for other dates when maps are exact.

PISCIS AUSTRINUS *(PICE-iss aw-STRY-nus)*

English Name: the southern fish

Size: small, 60th largest

When to Look: visible in the late evening sky from August through October

Notes: Easy to spot because of its bright star Fomalhaut; the rest of the constellation is dim and unremarkable. Fomalhaut is the most southerly first-magnitude star visible from most of our region and also the only first-magnitude star in the northern skies during the fall season.

WHERE TO LOOK: HORIZON GRAPH

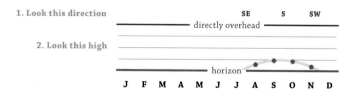

1. Look this direction

2. Look this high

Shown 11pm DST/10pm ST on the 15th. If viewing earlier/later, adjust 1 month for every 2 hours. (At 9pm in Sep, use Aug; at 1am in Sep, use Oct.) See page 10 for more info.

Stars in Piscis Austrinus

	LIGHT YEARS	MAGNITUDE	NAME ORIGIN
Fomalhaut (α)	22	1.16	Arabic for "fish's mouth"

Every star is ranked in terms of brightness, or magnitude. The brighter the star, the lower the magnitude. Stars with negative magnitudes are very bright. The sixth magnitude is the limit of human vision; the brightest stars are first magnitude or less.

The constellation contains only one star brighter than fourth magnitude, the first-magnitude star Fomalhaut.

Fomalhaut (α): The only bright star in the constellation and the 13th-brightest star in our sky. Its name is Arabic for "fish's mouth." It is the only first-magnitude star visible in this part of the skies and is sometimes called the Lonely Star of Autumn. It is also the most southerly first-magnitude star that can be seen from most of our region. The star is a close neighbor to our sun, lying just 22 light years from Earth. In 2008, Fomalhaut became the first star besides our sun to have an orbiting planet detected in a photograph. The images, taken with the Hubble Space Telescope, revealed a planet about the size of Jupiter in an orbit nearly four times larger than the orbit of Neptune. At this distance, the planet, named Fomalhaut b, takes 872 Earth years to make a single orbit around its star.

Mythology/History:

Prior to the twentieth century, this constellation was called Piscis Notius, which means "Great Fish." It is an ancient constellation that predates Greek civilization. The Greek mythology surrounding the constellation is somewhat obscure and set in the Middle East, which strongly suggests a Babylonian origin.

According to Greek writers, the Great Fish lived in a lake in Syria together with its two offspring, sometimes said to be represented by the constellation Pisces. One night Aphrodite's daughter Derceto, the Syrian Goddess of Fertility, fell into the lake and was rescued by the Great Fish.

Grateful for her safe return, Derceto placed the images of the fish in the stars and declared the animals to be sacred. From that time on, many Syrians abstained from eating fish and paid tribute by including golden and silver images of fish as household guardians. Temples built to honor Derceto were filled with fish ponds and their priests would eat fish as part of their daily sacrament.

Piscis Austrinus is usually depicted lying on its back, drinking the water that is spilling out of the urn carried by Aquarius.

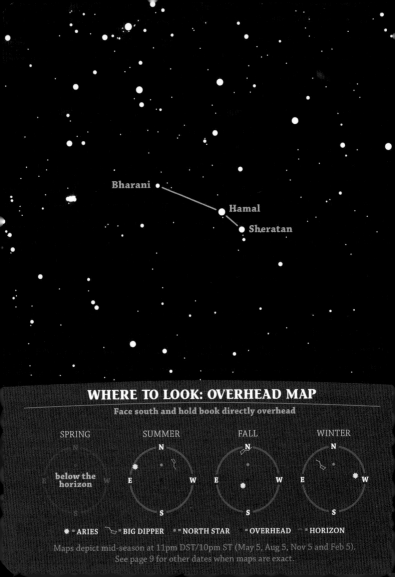

Bharani

Hamal

Sheratan

WHERE TO LOOK: OVERHEAD MAP

Face south and hold book directly overhead

SPRING

below the horizon

SUMMER

N
E W
S

FALL

N
E W
S

WINTER

N
E W
S

✳ = ARIES ⟍⟋ = BIG DIPPER •• = NORTH STAR • = OVERHEAD ⟍ = HORIZON

Maps depict mid-season at 11pm DST/10pm ST (May 5, Aug 5, Nov 5 and Feb 5).
See page 9 for other dates when maps are exact.

ARIES (AIR-eez)

English Name: the ram

Size: medium, 39th largest

When to Look: most prominent from October through January and visible in the late evening sky from September through February

Notes: Though small and somewhat faint, this constellation was very important historically. Around 1800 BC, the first day of spring occurred when the sun was in Aries. Although the sun is now in Pisces on the first day of spring, the location in the sky where the sun crosses the equator on its journey north is still called the First Point of Aries.

zodiac

WHERE TO LOOK: HORIZON GRAPH

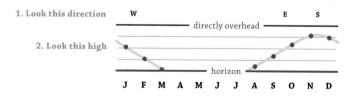

1. Look this direction

2. Look this high

W   E   S

directly overhead

horizon

J F M A M J J A S O N D

Shown 11pm DST/10pm ST on the 15th. If viewing earlier/later, adjust 1 month for every 2 hours. (At 9pm in Sep, use Aug; at 1am in Sep, use Oct.) See page 10 for more info.

Stars in Aries

	LIGHT YEARS	MAGNITUDE	NAME ORIGIN
Hamal (α)	66	2.0	Arabic for "lamb"
Sheratan (β)	66	2.6	Arabic for "pair"
Bharani (41 Arietis)	159	3.6	Hindu goddess of good luck

Every star is ranked in terms of brightness, or magnitude. The brighter the star, the lower the magnitude. Stars with negative magnitudes are very bright. The sixth magnitude is the limit of human vision; the brightest stars are first magnitude or less.

Hamal (α): The brightest star in the constellation, its name means "lamb" and is the Arabic name for the whole constellation.

Sheratan (β): The name means "a pair" in Arabic and is thought to refer to the ram's two horns.

Bharani (41 Arietis): The third-brightest star in the constellation, it is named after the Hindu goddess of good luck. The star lacks a Bayer Greek letter designation. At the time Bayer was assigning Greek letters to the stars, this star belonged to a constellation called *Musca Borealis*, Latin for "the Northern Fly," where it was labeled the alpha star. This constellation is no longer recognized and Bharani was never assigned a new Greek letter.

Ecliptic: The sun is in Aries from April 17 to May 11. When the zodiac was created 2,500 years ago, the sun was in Aries on the first day of spring. For this reason, it is the first constellation in the zodiac, and its astrological period runs from March 21 to April 19.

Mythology/History:

In Greek mythology, this constellation represents a ram with golden fleece sent by Zeus to save two children from sacrifice. Phrixus and Helle were the children of King Athamas. Their step-mother, Ino, detested them and plotted to kill them. She tricked the farmers into planting sterile grain, causing the crop to fail. At the time, it was common to sacrifice the king's children to the gods in time of famine. Seeing the deceit, Zeus sent a magical ram with golden fleece to fly the children to safety in the east. While flying over the sea, Helle lost her grip and fell, drowning in the waters that separate Europe from Asia. The Greeks called this place Hellespont in remembrance. Phrixus was delivered safely to Colchis on the Black Sea. Upon arriving, the ram gave its fleece to Colchis's king, Æetes, before flying up to take a place in the stars. It is said that the reason Aries is so faint is because the ram left his golden fleece behind on Earth.

Later, the golden fleece figured into the story of Jason and the Argonauts. Jason's stepbrother, Pelias, usurped the throne and imprisoned Jason's father. Pelias promised to make Jason king if he could obtain the fleece, which was guarded by a sleepless dragon. In one of the most famous Greek stories, Jason and a band of heroes retrieved the fleece. The astrological symbol for Aries (♈) represents the horns of the ram.

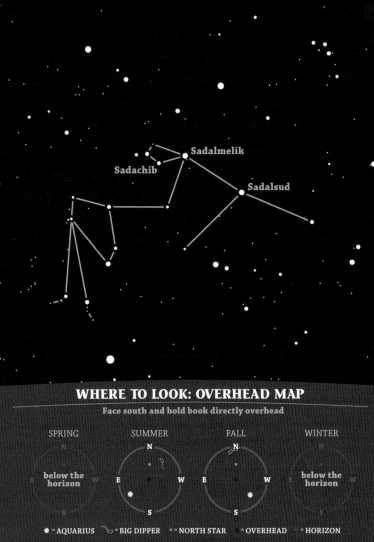

Sadachib

Sadalmelik

Sadalsud

WHERE TO LOOK: OVERHEAD MAP

Face south and hold book directly overhead

SPRING

below the horizon

SUMMER

N

E · W

S

FALL

N

E · W

S

WINTER

below the horizon

✹ = AQUARIUS   〰 = BIG DIPPER   • = NORTH STAR   • = OVERHEAD   ◯ = HORIZON

Maps depict mid-season at 11pm DST/10pm ST (May 5, Aug 5, Nov 5 and Feb 5).
See page 9 for other dates when maps are exact.

AQUARIUS *(uh-QUAIR-ee-us)*

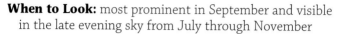

zodiac

English Name: the water bearer

Size: large, 10th largest

When to Look: most prominent in September and visible in the late evening sky from July through November

Notes: The second-largest constellation in the zodiac, Aquarius is comprised of relatively faint stars spilling across a large section of the autumn sky. Its large size makes it fairly easy to locate, but its complex shape and relatively dim stars make it a tricky, but rewarding, constellation to trace.

WHERE TO LOOK: HORIZON GRAPH

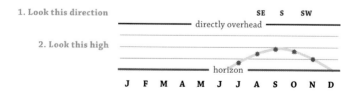

1. Look this direction
— directly overhead —
SE S SW

2. Look this high

horizon

J F M A M J J A S O N D

Shown 11pm DST/10pm ST on the 15th. If viewing earlier/later, adjust 1 month for every 2 hours. (At 9pm in Sep, use Aug; at 1am in Sep, use Oct.) See page 10 for more info.

Stars in Aquarius

	LIGHT YEARS	MAGNITUDE	NAME ORIGIN
Sadalsud (β)	600	2.9	Arabic for "luckiest of the lucky"
Sadalmelik (α)	750	3.0	Arabic for "lucky one of the king"
Sadachib (γ)	160	3.8	Arabic for "lucky star of the tent"

Every star is ranked in terms of brightness, or magnitude. The brighter the star, the lower the magnitude. Stars with negative magnitudes are very bright. The sixth magnitude is the limit of human vision; the brightest stars are first magnitude or less.

Sadalsud (β): The brightest star in the constellation. Its name means "luckiest of the lucky" in Arabic.

Sadalmelik (α): Arabic for "lucky one of the king."

Sadachib (γ): Arabic for "lucky star of the tent." It is part of a small Arabic constellation called "the tent" that is recognized in many European depictions as the urn from which Ganymede pours.

Ecliptic: The sun passes through Aquarius from February 14 to March 9. The eleventh constellation in the zodiac, Aquarius's astrological period runs from January 20 to February 18.

Mythology/History:

Located in the area of the sky the ancient Greeks referred to as "the sea," Aquarius is considered the source of all the celestial waters. This important constellation has been associated with water since antiquity, when the sun was in Aquarius at the beginning of the rainy season. An ancient Sumerian myth even names this constellation as the source of the waters that gave rise to the Great Flood. The zodiac symbol for Aquarius (♒) is the Egyptian hieroglyph for water.

In Greek mythology, the constellation is said to be the image of Ganymede, the young shepherd boy who was considered to be the most beautiful child ever born to mortals. Zeus took the boy to Mount Olympus to serve as cupbearer to the gods. In thanks for his service, Ganymede was granted immortality and placed among the stars.

More recently, Aquarius was made famous in the 1967 musical *Hair*, which proclaimed the dawning of the Age of Aquarius—a proclamation about 600 years too early. Astrological ages are named after the constellation that the sun is in on the first day of spring, and each lasts about 2,150 years. In about 600 years, the sun will enter Aquarius on the first day of spring, marking the real beginning of the Age of Aquarius.

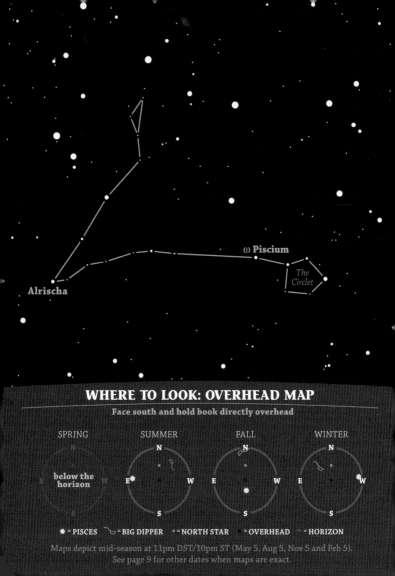

ω Piscium

The Circlet

Alrischa

WHERE TO LOOK: OVERHEAD MAP

Face south and hold book directly overhead

SPRING

below the horizon

SUMMER

N
E — W
S

FALL

N
E — W
S

WINTER

N
E — W
S

✷ = PISCES ⌐= BIG DIPPER •= NORTH STAR ▪ = OVERHEAD ⌐= HORIZON

Maps depict mid-season at 11pm DST/10pm ST (May 5, Aug 5, Nov 5 and Feb 5).
See page 9 for other dates when maps are exact.

PISCES *(PICE-eez)*

English Name: the fishes

Size: large, 14th largest

When to Look: most prominent in October and November and visible in the late evening sky from August through January

Notes: A large, faint constellation that wraps around two sides of the Great Square of Pegasus. It contains no stars brighter than of the fourth magnitude, making it tricky to spot and even harder to trace. The easiest part of the constellation to see is the "circlet," which represents one of the two fishes and is located directly below the Great Square. One of the constellations of the zodiac, its southern extension lies along the ecliptic and is often home to the moon and planets.

zodiac

WHERE TO LOOK: HORIZON GRAPH

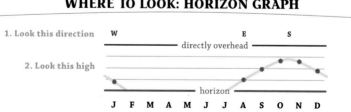

1. Look this direction

2. Look this high

Shown 11pm DST/10pm ST on the 15th. If viewing earlier/later, adjust 1 month for every 2 hours. (At 9pm in Sep, use Aug; at 1am in Sep, use Oct.) See page 10 for more info.

Stars in Pisces

	LIGHT YEARS	MAGNITUDE	NAME ORIGIN
Alrischa (α)	140	3.8	Arabic for "the knot"

Every star is ranked in terms of brightness, or magnitude. The brighter the star, the lower the magnitude. Stars with negative magnitudes are very bright. The sixth magnitude is the limit of human vision; the brightest stars are first magnitude or less.

Alrischa (α): This star marks the point where the lines to the two fish meet. Fittingly, its name is Arabic for "the knot."

The Circlet: A ring of 5 fourth- and fifth-magnitude stars that forms the body of the southern fish. It sits just south of the Great Square of Pegasus.

First Point of Aries: This point marks the intersection of the ecliptic and the celestial equator—the point where the sun crosses the equator on its journey north marking the first day of spring. When the zodiac was created, this point was in the constellation Aries. Although this point now occurs in the constellation Pisces (because of the Earth's wobble on its axis, which is called procession), it retains its anachronistic name. The point is most clearly marked by ω Piscium, which lies just north of the celestial equator.

Ecliptic: The sun passes through Pisces from March 10 to April 16, residing in the constellation on the first day of spring. The twelfth constellation in the zodiac, its astrological period runs from February 19 to March 20.

Mythology/History:

Pisces is an ancient constellation that has its origins in the Babylonian zodiac. There are two main myths associated with this constellation; both occur in the Middle East. In one myth, these two fish are the offspring of the Great Fish, Piscis Austrinus (page 189).

In the other myth, they are said to be the fish the gods Aphrodite and her son Eros transformed themselves into to escape from the monster Typhon. Typhon was the son of Gaia, the goddess of the Earth, and Tartarus, the pit of hell. Typhon was the largest and most hideous creature ever to live, with a hundred serpents' heads and a hundred serpents' tails. He waged war on the gods of Olympus and tried to replace Zeus as king. Some writers say that Typhon stole the thunderbolts of Zeus and brought storms that ravaged the land. In the chaos, most of the Olympian gods fled. In Syria, Aphrodite, the goddess of love and beauty, and her son Eros (Cupid), transformed themselves into fish to escape. Many Syrians abstained from eating fish out of fear that they would be consuming the gods. Zeus confronted Typhon and a tremendous battle ensued, causing earthquakes and tsunamis. Zeus defeated Typhon, imprisoning him under Mount Etna where he's the source of that mountain's eruptions.

The zodiac symbol for Pisces (♓) represents the curved bodies of two fish joined together.

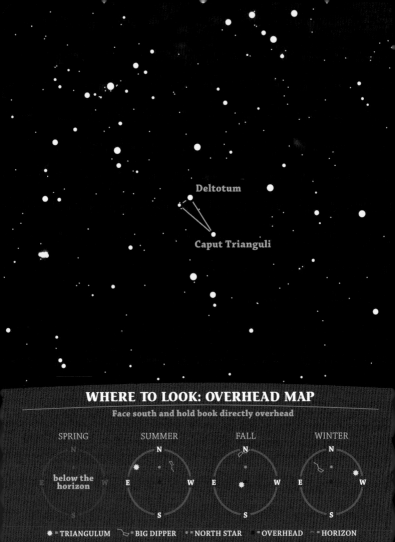

Deltotum

Caput Trianguli

WHERE TO LOOK: OVERHEAD MAP

Face south and hold book directly overhead

SPRING

below the horizon

SUMMER

FALL

WINTER

✳ = TRIANGULUM   〰 = BIG DIPPER   • = NORTH STAR   ▪ = OVERHEAD   ⎯ = HORIZON

Maps depict mid-season at 11pm DST/10pm ST (May 5, Aug 5, Nov 5 and Feb 5).
See page 9 for other dates when maps are exact.

TRIANGULUM *(try-ANG-gyuh-lum)*

English Name: the triangle

Size: very small, 78th largest

When to Look: most prominent from October through January and visible in the late evening sky from August through February

Notes: A very small constellation made up primarily of three moderately bright stars forming a narrow triangle. Though unremarkable, the constellation has been known since well before Greek times. Moderately difficult to spot, but its compact shape makes it very easy to trace.

WHERE TO LOOK: HORIZON GRAPH

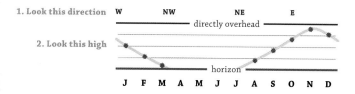

1. Look this direction

2. Look this high

Shown 11pm DST/10pm ST on the 15th. If viewing earlier/later, adjust 1 month for every 2 hours. (At 9pm in Sep, use Aug; at 1am in Sep, use Oct.) See page 10 for more info.

Stars in Triangulum

	LIGHT YEARS	MAGNITUDE	NAME ORIGIN
Deltotum (β)	124	3.0	Reference to Greek letter delta
Caput Trianguli (α)	59	3.4	Latin for "head of the triangle"

Every star is ranked in terms of brightness, or magnitude. The brighter the star, the lower the magnitude. Stars with negative magnitudes are very bright. The sixth magnitude is the limit of human vision; the brightest stars are first magnitude or less.

Deltotum (β): The brightest star in the constellation, its name refers to the Greek letter delta (Δ) and was an alternate name for the constellation in ancient times.

Caput Trianguli (α): The constellation's second-brightest star, marking the tip of the triangle. Its name is Latin for "head of the triangle." Even more rarely it is called *mothallah*, which is Arabic for "triangle."

Mythology/History:

This constellation is ancient but has no mythology clearly associated with it. Some ancient writers suggest that the stars point to the faint stars of Aries. Others said that it represents the capital letter delta (Δ), honoring the first letter of Zeus's name in Greek. Others called the constellation Deltotum and associated it with the delta of the Nile.

The most interesting story in the ancient Greek writings, however, is one relayed by Eratosthenes, a contemporary of Ptolemy. He suggested that Egypt was modeled after these stars—with the Nile carving the land into a delta shape thereby protecting it and making it fertile. This account is unique among Greek stories, as in all other Greek myths the configuration of the stars is the result of events on Earth. In this story, the shape of the Earth itself is said to be a result of the stars. While unique in Greek mythology, ancient Egyptian lore held that Egypt was a reflection of the heavens—suggesting that this tiny constellation may be quite old.

The Romans held that these stars represented the island of Sicily, which was placed in the heavens by Jupiter (Zeus) at the request of Ceres, the patron goddess of Sicily and the goddess of agriculture. In 1801, the constellation was the site of the very first asteroid to be discovered. By coincidence, the discovery was made from an observatory in Sicily.

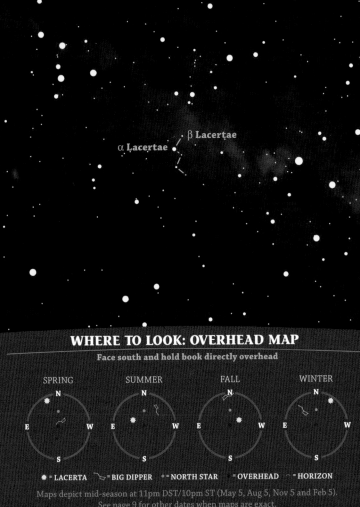

β Lacertae

α Lacertae

WHERE TO LOOK: OVERHEAD MAP

Face south and hold book directly overhead

SPRING   SUMMER   FALL   WINTER

✳ = LACERTA   ⌐ = BIG DIPPER   • = NORTH STAR   = OVERHEAD   ⌒ = HORIZON

Maps depict mid-season at 11pm DST/10pm ST (May 5, Aug 5, Nov 5 and Feb 5).
See page 9 for other dates when maps are exact.

LACERTA *(luh-SER-tuh)*

English Name: the lizard

Size: small, 68th largest

When to Look: most prominent from August through November and visible in the late evening sky from June through January

Notes: A dim, modern constellation located in between Andromeda, Cassiopeia and Cygnus. Though it is quite small and contains no bright stars, its position on the edge of the Milky Way makes it an interesting area to scan with binoculars. The main stars of the constellation form a small, dim "W," which is sometimes referred to as the "Little Cassiopeia" because of its resemblance to its bright neighbor.

WHERE TO LOOK: HORIZON GRAPH

1. Look this direction

NW   N   NE   N   NW

directly overhead

2. Look this high

horizon

J F M A M J J A S O N D

Shown 11pm DST/10pm ST on the 15th. If viewing earlier/later, adjust 1 month for every 2 hours. (At 9pm in Sep, use Aug; at 1am in Sep, use Oct.) See page 10 for more info.

Stars in Lacerta

	LIGHT YEARS	MAGNITUDE	NAME ORIGIN
α **Lacertae**	98	3.8	No common name
β **Lacertae**	215	4.4	No common name

Every star is ranked in terms of brightness, or magnitude. The brighter the star, the lower the magnitude. Stars with negative magnitudes are very bright. The sixth magnitude is the limit of human vision; the brightest stars are first magnitude or less.

While the constellation is small and contains no stars brighter than of the fourth magnitude, the north end of Lacerta sits in the Milky Way and is an interesting area to scan with binoculars. None of the stars in the constellation have common names.

α **Lacertae**: The brightest star in the constellation. Unspectacular at magnitude 3.8, but it is backed by the Milky Way.

β **Lacertae**: The only other star in the constellation with a Bayer designation—it was given a Greek letter despite being only the fourth-brightest star in the constellation. (Greek letters are generally assigned in order of brightness.)

Mythology/History:

A modern constellation created by Johannes Hevelius to fill an unformed area of the sky in between the constellations Cygnus and Andromeda.

Hevelius explained that because this area of the sky is not very large he placed a lizard here since nothing else would fit. About a decade earlier, French astronomer Augustin Royer had named this region of the sky *Sceptrum et Manus Iustitiae*, Latin for "the scepter and hand of justice," in honor of King Louis XIV of France. Due in large part to its awkward name, Royer's constellation was largely ignored by other astronomers.

Nearly a century later, German astronomer Johann Bode further divided this area of the sky and created a new constellation at the junction of Andromeda, Cassiopeia and Lacerta called *Frederici Honores* (the Glory of Frederick) in honor of Frederick the Great of Prussia.

When the International Astronomical Union created its official list of constellations, only Hevelius's contribution to this region of the sky was recognized. Six of his other creations were also recognized: Canes Venatici, Leo Minor, Lynx, Scutum, Sextans and Vulpecula.

α **Fornacis**

β **Fornacis**

WHERE TO LOOK: OVERHEAD MAP

Face south and hold book directly overhead

SPRING

SUMMER

FALL

WINTER

below the horizon

below the horizon

below the horizon

* = FORNAX   = BIG DIPPER   • = NORTH STAR   • = OVERHEAD   = HORIZON

Maps depict mid-season at 11pm DST/10pm ST (May 5, Aug 5, Nov 5 and Feb 5).
See page 9 for other dates when maps are exact.

FORNAX (FOR-naks)

English Name: the furnace

Size: small, 41st largest

When to Look: visible in the late evening sky from November through December

Notes: A modern constellation published in 1751 by Nicolas Louis de Lacaille. Lacaille's original name was *Fornax Chemica*, Latin for "the Chemical Furnace." Dim and uninteresting, with only two stars of fourth magnitude and no named stars, it lies in an undistinguished area of the sky.

WHERE TO LOOK: HORIZON GRAPH

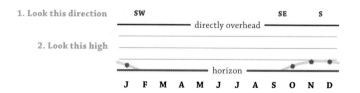

1. Look this direction

SW   SE   S
———————————— directly overhead ————————————

2. Look this high

———————————————— horizon ————————————————

J F M A M J J A S O N D

Shown 11pm DST/10pm ST on the 15th. If viewing earlier/later, adjust 1 month for every 2 hours. (At 8pm in Nov, use Oct; at 12am in Nov, use Dec.) See page 10 for more info.

Stars in Fornax

	LIGHT YEARS	MAGNITUDE	NAME ORIGIN
α **Fornacis**	46	3.9	No common name
β **Fornacis**	200	4.4	No common name

Every star is ranked in terms of brightness, or magnitude. The brighter the star, the lower the magnitude. Stars with negative magnitudes are very bright. The sixth magnitude is the limit of human vision; the brightest stars are first magnitude or less.

Fornax contains only two stars of the fourth magnitude and little else of interest to those viewing the constellations with the naked eye or through binoculars. Its major claim to fame is a group of galaxies that can be seen with a telescope in the constellation's southeast corner.

α **Fornacis**: The constellation's brightest star at magnitude 3.9, it has no common name and lies 46 light years from Earth.

β **Fornacis**: The second-brightest star in the constellation, barely qualifying for the fourth magnitude.

Mythology/History:

Fornax is a modern constellation created by the French astronomer Nicolas Louis de Lacaille in 1756. Lacaille's most important contributions to astronomy were his extensive star charts of the southern skies. Lacaille produced a catalog of over 10,000 stars and defined 14 new constellations.

Fornax is the Latin word for "furnace." It is also the name of the Roman goddess of furnaces and ovens, who is also the patron of bakers. Lacaille's creation, however, is not a reference to either the goddess or the baker's furnace. Originally named *Fornax Chemica* (Latin for "the Chemical Furnace"), this constellation honored the small burners used by chemists for distillation.

Distillation has a history going back at least 2,000 years, since alchemists in Spain began using the process to isolate mercury from its ore, cinnabar. The distillation process involves heating a mixture to boiling in order to separate it into its component parts. As a mixture is heated to boiling, its different components will boil off at different temperatures. By capturing these vapors, it's possible to obtain a pure compound. By 1000 AD, distillation was being used to make brandy from wine, and it has since helped isolate many important chemicals and is used in manufacture of petroleum products and alcohol.

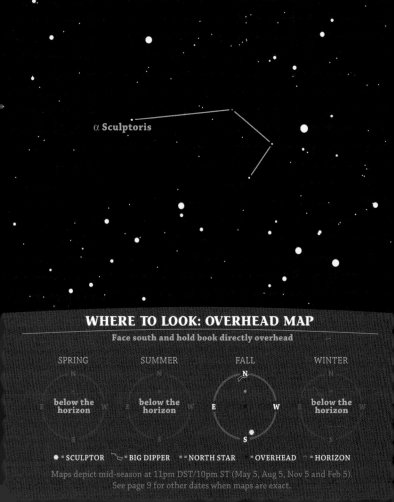

α **Sculptoris**

WHERE TO LOOK: OVERHEAD MAP

Face south and hold book directly overhead

SPRING   SUMMER   FALL   WINTER

below the horizon   **below the horizon**   **below the horizon**

✳ = SCULPTOR   ⌁ = BIG DIPPER   ▪▪ = NORTH STAR   ▫ = OVERHEAD   = HORIZON

Maps depict mid-season at 11pm DST/10pm ST (May 5, Aug 5, Nov 5 and Feb 5).
See page 9 for other dates when maps are exact.

SCULPTOR *(SCULP-ter)*

English Name: the sculptor

Size: small, 36th largest

When to Look: visible in the late evening sky from October through November

Notes: A dim, modern constellation. It lies in the direction of the south galactic pole and provides one of the clearest views out of our galaxy. It looks nothing like its namesake. Although it is fully visible throughout most of our range, its dim stars are very difficult to see close to the horizon.

WHERE TO LOOK: HORIZON GRAPH

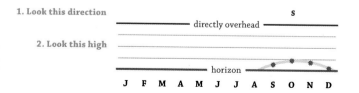

1. Look this direction

S

—— directly overhead ——

2. Look this high

—— horizon ——

J F M A M J J A S O N D

Shown 11pm DST/10pm ST on the 15th. If viewing earlier/later, adjust 1 month for every 2 hours. (At 9pm in Oct, use Sep; at 1am in Oct, use Nov.) See page 10 for more info.

Stars in Sculptor

	LIGHT YEARS	MAGNITUDE	NAME ORIGIN
α **Sculptoris**	672	4.3	No common name

Every star is ranked in terms of brightness, or magnitude. The brighter the star, the lower the magnitude. Stars with negative magnitudes are very bright. The sixth magnitude is the limit of human vision; the brightest stars are first magnitude or less.

None of the stars in the constellation have common names and the brightest stars are only fourth magnitude. The constellation also contains little of interest for viewers with binoculars.

α **Sculptoris**: The brightest star in the constellation, with a magnitude of just 4.3.

South Galactic Pole: When we look toward Sculptor, we are looking perpendicular to the disk of our own galaxy. With relatively few stars and little interstellar dust in the way, we get one of our clearest views of deep space. The Sculptor Galaxy group lies in this region of space; it is one of the groups of galaxies closest to us. At a distance of about 12 million light years, these far-off galaxies are too faint to see with the naked eye or binoculars.

Mythology/History:

This constellation was created by French astronomer Nicolas Louis de Lacaille in the mid-eighteenth century. Like most of the 14 constellations that de Lacaille defined, it is very faint and looks nothing like its namesake. Unlike most of de Lacaille's creations, it is not named after a specific instrument of art or science, but rather commemorates the sculptor's studio. De Lacaille's original name for the constellation, shown on his 1756 star chart, was *l'Atelier du Sculpteur*, which is French for "the Sculptor's Studio." A few years later, he published another star chart showing the constellation with the Latinized name *Apparatus Sculptoris*.

De Lacaille's original drawings of the constellation showed a block of marble sitting on top of a small table being carved into a bust, and a second table with a small collection of sculptor's tools. The drawing had little to do with the stars in the constellation. Most later depictions of the constellation showed a single small table with a carved bust and tools sitting on top of it. When the International Astronomical Union defined the 88 modern constellations, they dispensed with the studio entirely and shortened the name of de Lacaille's creation to Sculptor.

THE WINTER SKY
(overhead, facing south)

This map shows the winter sky as it appears at 12am on Jan 5, 10pm on Feb 5, and 8pm on Mar 5. For other times this chart can be used, see page 9.

IN THE WINTER, the dark side of the Earth faces toward the outer arms of the Milky Way, which are filled with stellar nurseries and bright, hot, young stars. Fully half of the brightest stars in our sky are in the winter constellations. The crisp, quiet winter nights make the stars seem to glisten even more than the rest of the year. If you can brave the cold, the long winter nights are a magical time to stargaze.

Orion (page 223) is the brightest of all the constellations and dominates the winter sky. Look here for the brilliant blue Rigel and fiery red Betelgeuse, the bright trio of Orion's Belt, and the soft glow of the Orion Nebula. In Canis Major (page 227) you can see Sirius, the brightest star in the entire sky.

Winter is also a great time to see star clusters. Look in Taurus (page 231) for the magnificent Pleiades and Hyades star clusters; in Cancer (page 247) for the Praesepe, a star cluster; or scan the band of the Milky Way with a pair of binoculars.

The winter constellations are in blue on the seasonal sky map to the left.

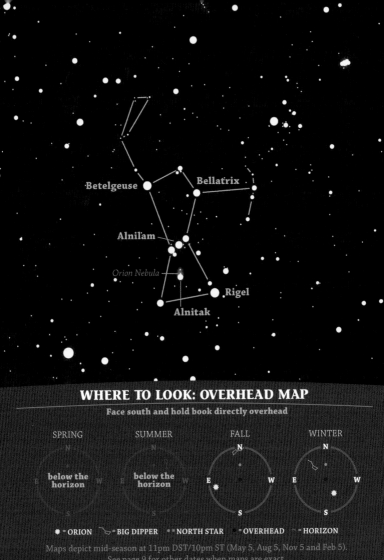

Betelgeuse

Bellatrix

Alnilam

Orion Nebula

Rigel

Alnitak

WHERE TO LOOK: OVERHEAD MAP

Face south and hold book directly overhead

SPRING	SUMMER	FALL	WINTER
N	N	N	N
below the horizon	**below the horizon**	E ... W	E ... W
S	S	S	S

✴ = ORION ◟ = BIG DIPPER ∗ = NORTH STAR ∗ = OVERHEAD ⌐ = HORIZON

Maps depict mid-season at 11pm DST/10pm ST (May 5, Aug 5, Nov 5 and Feb 5).
See page 9 for other dates when maps are exact.

ORION (oh-RYE-un)

English Name: the hunter

Size: medium, 26th largest

When to Look: most prominent in December and January and visible in the late evening sky from November through March

Notes: The most spectacular and prominent constellation in the sky, Orion contains more bright stars than any other constellation and is the only constellation that contains two first-magnitude stars. It also contains the sky's brightest stellar nebula. Sitting on the celestial equator, Orion is visible from everywhere on Earth.

WHERE TO LOOK: HORIZON GRAPH

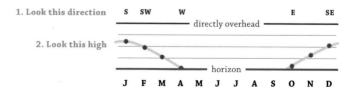

1. Look this direction   S   SW   W   E   SE

directly overhead

2. Look this high

horizon

J F M A M J J A S O N D

Shown 11pm DST/10pm ST on the 15th. If viewing earlier/later, adjust 1 month for every 2 hours. (At 8pm in Jan, use Dec; at 12am in Jan, use Feb.) See page 10 for more info.

223

Stars in Orion

	LIGHT YEARS	MAGNITUDE	NAME ORIGIN
Rigel (β)	900	0.2	Arabic for "foot"
Betelgeuse (α)	640	0.58	Arabic for "shoulder of the giant"
Alnitak (ζ)	820	1.5	Arabic for "the girdle"
Bellatrix (γ)	240	1.6	Latin for "warrioress"
Alnilam (ε)	1,200	1.7	Arabic for "string of pearls"

Every star is ranked in terms of brightness, or magnitude. The brighter the star, the lower the magnitude. Stars with negative magnitudes are very bright. The sixth magnitude is the limit of human vision; the brightest stars are first magnitude or less.

Rigel (β): The brightest star in the constellation and the fifth-brightest in our sky, Rigel is the most luminous star in our region of the galaxy. Historically, the star has been very important for navigation, as it is visible from every ocean in the world.

Betelgeuse (α): The seventh-brightest star in our sky, and the second-brightest in the constellation. With a red-orange glow, it makes a striking contrast with the brilliant blue-white Rigel. Betelgeuse is one of the largest known stars, with a diameter greater than the orbit of Mars.

Alnitak (ζ): The 16th-brightest star in our sky, Alnitak is actually a triple star system.

Orion Nebula (M42): A diffuse nebula clearly visible to the naked eye under good conditions. One of the most closely studied celestial objects, the nebula has taught astronomers much about stellar formation.

Orion's Belt: A bright asterism of three stars in a straight line, these stars have been known by many names the world over. Examples include: the Three Kings (Africa), the Three Marys (Latin America), and the String of Pearls (the Middle East).

Mythology/History:

Orion is one of only three constellations mentioned in the Bible. The name Orion and the association of these stars with a hunter appears to predate Greek civilization by thousands of years, dating back at least to the Akkadian Empire of Sumeria.

Although there is no central account of his life, stories about Orion are woven into many classical Greek legends. These stories include several versions of his birth and death, and connect him with many other constellations in the sky. One myth says that Orion was a persistent and unwanted suitor of the god Atlas's daughters, known as the Pleiades. At Atlas's request, Zeus transformed the sisters into stars to keep them safe. Orion is said to still be pursuing them across the sky.

In another myth, Orion boasted that he was such a great hunter that he was a match for any beast. The goddess Hera decided to teach him a lesson and sent a tiny scorpion to sting him. Orion crushed the creature with his club, but only after it had fatally stung him. The pair were both placed in the sky as a reminder of the cost of hubris. Several other constellations are connected with the hunter as well. Canis Major and Canis Minor are said to be his hunting dogs, while Taurus and Lepus are said to be his quarry.

Sirius

Murzim

Wenzen

Adhara

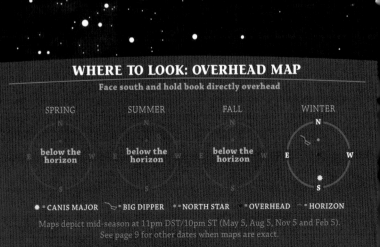

WHERE TO LOOK: OVERHEAD MAP

Face south and hold book directly overhead

SPRING	SUMMER	FALL	WINTER
below the horizon	**below the horizon**	**below the horizon**	

✳ = CANIS MAJOR ⌐ = BIG DIPPER ＊ = NORTH STAR ● = OVERHEAD ⌒ = HORIZON

Maps depict mid-season at 11pm DST/10pm ST (May 5, Aug 5, Nov 5 and Feb 5).
See page 9 for other dates when maps are exact.

CANIS MAJOR *(CANE-iss MAY-jer)*

English Name: the big dog

Size: medium, 25th largest

When to Look: most prominent in January and February and visible in the late evening sky from December through March

Notes: An ancient constellation filled with bright stars, including Sirius, the brightest star in the sky. Unmistakable and very easy to locate, though its somewhat indistinct shape and mix of bright and dim stars actually make it difficult to trace. Although the constellation is said to represent one of Orion's hunting dogs, most of the mythology associated with the constellation is actually about the star Sirius.

WHERE TO LOOK: HORIZON GRAPH

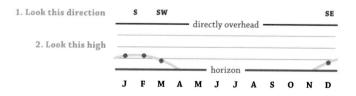

1. Look this direction

S SW SE

directly overhead

2. Look this high

horizon

J F M A M J J A S O N D

Shown 11pm DST/10pm ST on the 15th. If viewing earlier/later, adjust 1 month for every 2 hours. (At 8pm in Jan, use Dec; at 12am in Jan, use Feb.) See page 10 for more info.

Stars in Canis Major

	LIGHT YEARS	MAGNITUDE	NAME ORIGIN
Sirius (α)	8.6	–1.5	Greek for "searing"
Adhara (ε)	450	1.5	Arabic for "the virgins"
Wenzen (δ)	1800	1.8	Arabic for "weight"
Murzim (β)	500	1.98	Arabic for "the herald"

Every star is ranked in terms of brightness, or magnitude. The brighter the star, the lower the magnitude. Stars with negative magnitudes are very bright. The sixth magnitude is the limit of human vision; the brightest stars are first magnitude or less.

Sirius (α): The brightest star in the sky, Sirius is not an especially large or luminous star, but owes its brightness to the fact that it lies only 8.8 light years from Earth, making it the fifth-closest of all known stars. Because it is so bright, Sirius often flickers much more than other stars when it is close to the horizon. When watching the star, one can sometimes see flashes of blue, red and other vibrant colors as the star's light is refracted in Earth's atmosphere. Sirius is often referred to as "the dog star."

Adhara (ε): The brightest of the second-magnitude stars and the 22nd-brightest star in our sky, Adhara is nearly first magnitude.

Wenzen (δ): The third-brightest star in the constellation and the 27th brightest in our skies.

Murzim (β): The 33rd-brightest star in our skies, its common name refers to its position, rising just before Sirius and heralding the bright star's appearance.

The Milky Way passes between Sirius and Procyon in Canis Minor.

Mythology/History:

Home to Sirius, the brightest star in the sky, Canis Major is one of the oldest named constellations and has many myths associated with it.

Many classical texts associate the constellation with Laelaps, the dog that Zeus gave to his mistress Europa to protect her. Europa was a noblewoman from the Kingdom of Phoenicia whom Zeus seduced by taking the form of a white bull (see Taurus, page 233). Zeus gave Europa three gifts to protect her and provide for her: a bronze giant named Talos, who served as her guardian; a spear that never missed its target; and Laelaps, a hunting dog that could run down any game. After passing through a series of owners, Laelaps was brought to Thebes to hunt down the Teumessian fox, an animal sent by the gods to torment the countryside. The gods had made the fox so swift that it could never be caught. Laelaps gave chase to the fox, creating a paradox. Zeus saw the dilemma and ended the chase by turning the fox to stone and placing Laelaps in the stars.

Today, however, Canis Major is usually considered the hunting dog of Orion, which was placed in the sky after Orion was turned into a constellation.
There are no stories about his dog, however, and the only references that mention Orion even having a dog are those associated with constellations.

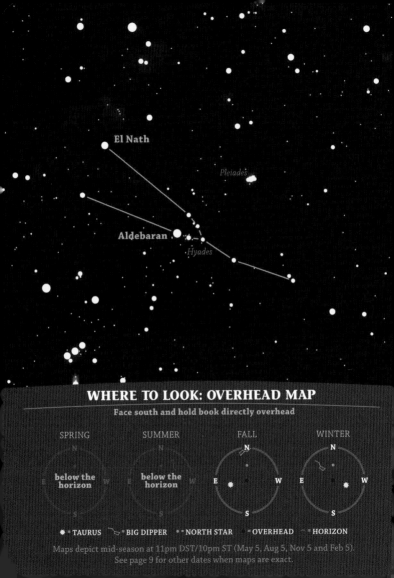

El Nath

Pleiades

Aldebaran

Hyades

WHERE TO LOOK: OVERHEAD MAP

Face south and hold book directly overhead

SPRING	SUMMER	FALL	WINTER

below the horizon **below the horizon**

✴ = TAURUS ↶ = BIG DIPPER • = NORTH STAR ▫ = OVERHEAD — = HORIZON

Maps depict mid-season at 11pm DST/10pm ST (May 5, Aug 5, Nov 5 and Feb 5).
See page 9 for other dates when maps are exact.

TAURUS *(TOR-us)*

zodiac

English Name: the bull

Size: large, 17th largest

When to Look: most prominent from November through January and visible in the late evening sky from October through March

Notes: A prominent and well-known constellation, it is home to the two most famous star clusters in the sky, the Pleiades and the Hyades. Easy to find because of its bright star, Aldebaran, it does resemble the head of a bull with long horns. One of the constellations of the zodiac, it lies along the ecliptic and is often home to the moon and planets.

WHERE TO LOOK: HORIZON GRAPH

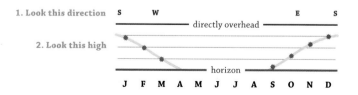

1. Look this direction

2. Look this high

Shown 11pm DST/10pm ST on the 15th. If viewing earlier/later, adjust 1 month for every 2 hours. (At 8pm in Jan, use Dec; at 12am in Jan, use Feb.) See page 10 for more info.

Stars in Taurus

	LIGHT YEARS	MAGNITUDE	NAME ORIGIN
Aldebaran (α)	65	0.9	Arabic for "follower"
El Nath (β)	131	1.7	Arabic for "butting one"

Every star is ranked in terms of brightness, or magnitude. The brighter the star, the lower the magnitude. Stars with negative magnitudes are very bright. The sixth magnitude is the limit of human vision; the brightest stars are first magnitude or less.

Aldebaran (α): The brightest star in the constellation and the ninth-brightest star in the sky, Aldebaran represents the eye of the bull. Its orange-red color can be seen with the naked eye.

El Nath (β): The constellation's second-brightest star and one of the 30 brightest stars in the sky.

Hyades: A large, loose star cluster located just west of Aldebaran. Known since classical times, the star cluster is named after the five sisters of Hyas, a great hunter who died tragically. The weeping sisters were placed in the heavens, where their tears became rain.

Pleiades (M45): The most famous cluster in the sky. Easily seen with the naked eye and spectacular in binoculars. Under good conditions, a person with very good eyesight can clearly see six stars clustered in a fuzzy patch about four times the diameter of the full moon. Binoculars clearly show the cluster's six brightest stars to form a tiny dipper. At one time, the cluster was known as a separate constellation. It is one of three constellations mentioned in the Bible (Orion and Ursa Major are the other two). Named for the daughters of Atlas and Pleione, legend says that the sisters were placed in the sky to protect them from the incessant courting of the hunter, Orion.

Ecliptic: The sun is located in Taurus from May 12 through June 18. The second constellation in the zodiac, its astrological period runs from April 20 to May 20.

Mythology/History:

It is likely Taurus was the very first zodiac constellation to be recognized, as it was associated with the start of spring in early human history. Taurus is associated with several Greek myths. It is sometimes said to represent the Cretan Bull that Hercules captured or the bull that sired the infamous Minotaur. Most commonly these stars are said to represent Zeus himself, disguised as a white bull to seduce the beautiful maiden Europa. Europa was a noblewoman from the kingdom of Phoenicia. Zeus fell in love with her but Europa's father kept her closely guarded and would not allow any suitors approach. Zeus transformed himself into a white bull and mingled in with her father's herds. While Europa was out on her daily walk, she approached the magnificent bull. Struck by its beauty and gentleness, she petted it and climbed onto its back. Zeus took this opportunity, ran down to the sea and swam to the island of Crete with Europa on his back. Once there, he revealed his true identity.

Zeus made Europa the first Queen of Crete and gave her three gifts: a bronze guardian, a spear that never missed and a hunting dog that could run down any game. She is considered the mother of Crete and the continent of Europe gets its name from her. The constellation shows only the bull's head and shoulders, as it depicts it swimming across the sea to Crete. Its astrological sign is ♉.

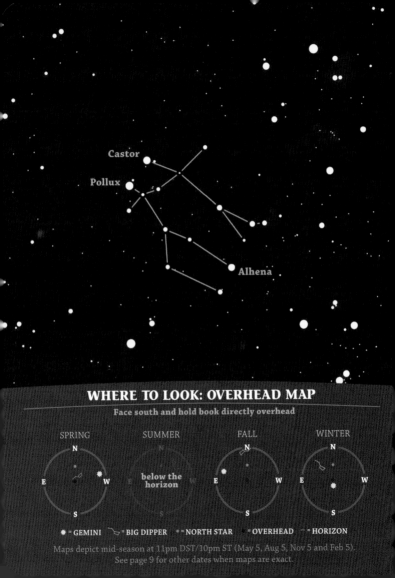

Castor

Pollux

Alhena

WHERE TO LOOK: OVERHEAD MAP

Face south and hold book directly overhead

SPRING	SUMMER	FALL	WINTER
	below the horizon		

✳ = GEMINI ⌐ = BIG DIPPER • = NORTH STAR ▪ = OVERHEAD ▬ = HORIZON

Maps depict mid-season at 11pm DST/10pm ST (May 5, Aug 5, Nov 5 and Feb 5).
See page 9 for other dates when maps are exact.

GEMINI *(JEM-uh-nye)*

zodiac

English Name: the twins

Size: medium, 30th largest

When to Look: most prominent from December through March and visible in the late evening sky from November through April

Notes: An easy constellation to locate because of its twin bright stars, Pollux and Castor, which mark the heads of the twins. Both the bright pair of stars and the entire constellation are referred to as "the twins." The ecliptic passes through the center of the constellation so the moon and planets can often be seen here.

WHERE TO LOOK: HORIZON GRAPH

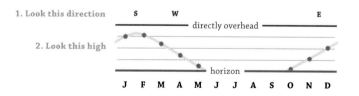

1. Look this direction

2. Look this high

Shown 11pm DST/10pm ST on the 15th. If viewing earlier/later, adjust 1 month for every 2 hours. (At 8pm in Jan, use Dec; at 12am in Jan, use Feb.) See page 10 for more info.

Stars in Gemini

	LIGHT YEARS	MAGNITUDE	NAME ORIGIN
Pollux (β)	36	1.1	Greek mythological character
Castor (α)	46	1.6	Greek mythological character
Alhena (γ)	105	1.9	Arabic for "mark on the neck of a camel"

Every star is ranked in terms of brightness, or magnitude. The brighter the star, the lower the magnitude. Stars with negative magnitudes are very bright. The sixth magnitude is the limit of human vision; the brightest stars are first magnitude or less.

Pollux (β): The brightest star in the constellation and the 12th-brightest star in the sky, Pollux also marks the northeast corner of the Great Hexagon.

Castor (α): The second-brightest second-magnitude star and the 18th-brightest star overall, Castor makes a worthy companion to its brighter twin, Pollux. What appears to us as a single star is actually a six-star system, though a telescope is needed to see its individual components.

Alhena (γ): A bright second-magnitude star backed by the Milky Way that marks one of the feet of the twins. Its unusual name refers to an obscure and forgotten Arab constellation.

Ecliptic: The sun passes through Gemini from June 19 to July 18 and resides among these stars on the summer solstice. The third constellation in the zodiac, its astrological period runs from May 21 to June 20.

The Milky Way passes through the feet of the twins.

Mythology/History:

Gemini is named after the twin brothers Castor and Pollux (sometimes referred to by his Greek name, Polydeuces). In Greek, the pair are called the *Dioscuri*, meaning "sons of Zeus." In Latin, they are simply referred to as *Gemini*, or "the twins."

Castor and Pollux were born after Zeus consorted with their mother Leda, the Queen of Sparta. Zeus became enamored with Leda, but she remained loyal to her husband. To deceive Leda, Zeus transformed into a swan and approached. Leda allowed the swan to sit on her lap as she rested. When she fell asleep, Zeus consorted with her. Zeus then flew off to the heavens, placing the image of the swan among the stars (see page 95). From this union, Leda produced two eggs. Helen of Troy hatched from one egg. The twins Pollux and Castor hatched from the other; Pollux was born immortal, Castor mortal.

Pollux and Castor were members of the Argonauts, Jason's crew in his quest for the golden fleece (page 195). According to legend, they were fine sailors and horsemen and the patrons of mariners and cavalry men. When Castor was killed, Pollux told Zeus that he didn't want to be immortal if his brother wasn't. Zeus permitted Pollux to give half of his immortality to Castor, allowing the twins to divide their time between the realm of the gods and the underworld. The zodiac symbol for Gemini is (♊), the Roman numeral 2.

Capella

Almaaz

El Nath

WHERE TO LOOK: OVERHEAD MAP

Face south and hold book directly overhead

SPRING

SUMMER

below the horizon

FALL

WINTER

✳ = AURIGA   ⌐ = BIG DIPPER   • = NORTH STAR   ■ = OVERHEAD   ◇ = HORIZON

Maps depict mid-season at 11pm DST/10pm ST (May 5, Aug 5, Nov 5 and Feb 5).
See page 9 for other dates when maps are exact.

AURIGA *(aw-RYE-guh)*

English Name: the charioteer

Size: medium, 21st largest

When to Look: most prominent from December through March and visible in the late evening sky from October through April

Notes: One of the most ancient constellations in the sky. Easy to locate because of its bright star, Capella, it is also easy to trace. Sitting across the Milky Way, it is a rich area of the sky to observe, especially with binoculars.

WHERE TO LOOK: HORIZON GRAPH

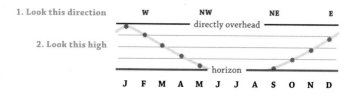

1. Look this direction

2. Look this high

Shown 11pm DST/10pm ST on the 15th. If viewing earlier/later, adjust 1 month for every 2 hours. (At 8pm in Jan, use Dec; at 12am in Jan, use Feb.) See page 10 for more info.

Stars in Auriga

	LIGHT YEARS	MAGNITUDE	NAME ORIGIN
Capella (α)	42	0.08	Latin for "little goat"
El Nath (β Tauri)	131	1.7	Arabic for "the butting one"
Almaaz (ε)	2,000	2.9*	Arabic for "goat"

Every star is ranked in terms of brightness, or magnitude. The brighter the star, the lower the magnitude. Stars with negative magnitudes are very bright. The sixth magnitude is the limit of human vision; the brightest stars are first magnitude or less.

* Drops to 3.8 once every 27 years

Capella (α): Capella is the fourth-brightest star in our sky and the northernmost first-magnitude star. Circumpolar above 45 degrees north latitude, it can be seen at least part of every night of the year from most of the United States.

El Nath (β Tauri): Once labeled as γ Aurigae but now officially part of the constellation Taurus.

Almaaz (ε): One of the most luminous stars in our region of the galaxy, it is quite bright despite its great distance. It is also an unusual variable; the star system includes a brilliant white supergiant that is eclipsed by a companion once every 27 years. During the eclipse, which lasts nearly 2 years, the star drops from magnitude 2.9 to 3.8. Remarkably, the primary star is a supergiant some 100 times the diameter of our sun and it is being eclipsed by something much larger than itself. It is still a mystery what exactly obscures this massive star from us. The next eclipse will run from 2009 to 2011. The event is of great interest to astronomers and many telescopes will be aimed at the star system to try to unlock its secrets.

The Milky Way passes through the constellation's center. This region contains several star clusters visible in binoculars.

Mythology/History:

One of the most ancient constellations in the sky, it was first identified as a charioteer by the Babylonians about 4,000 years ago.

In classical Greek mythology, the charioteer is usually identified as Erichthonius, son of Hephaestus and Gaia. Hephaestus was the blacksmith of the gods and Gaia was the goddess of the Earth. Erichthonius's name means "Earth-born." Like most characters in Greek mythology that were born from the Earth, Erichthonius is associated with serpents and was described as having the body of a serpent from his waist down.

When young, Erichthonius imitated the sun god, Helios, by building a chariot and harnessing four horses to it. (In Greek mythology, Helios crossed the sky each day from east to west in the chariot of the sun.) Zeus marveled at the youth's ingenuity and courage and honored him by placing him among the stars.

The constellation's bright star, Capella, is said to be the goat Amalthea that suckled the infant Zeus. When Zeus was preparing to go to fight the Titans, the Oracle told him to wear the skin of a goat as a shield. He took Amalthea's skin and, after returning victorious, covered the goat's bones with new skin and granted her immortality by placing her in the stars.

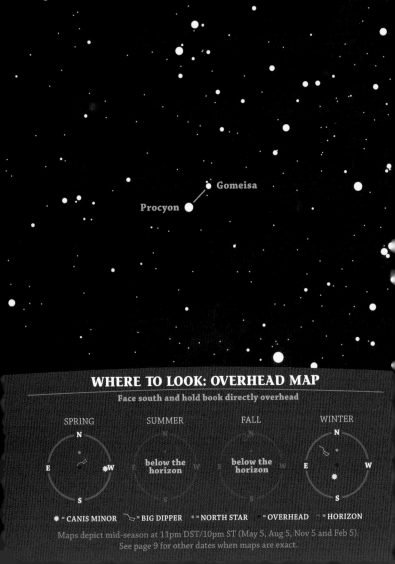

Gomeisa

Procyon

WHERE TO LOOK: OVERHEAD MAP

Face south and hold book directly overhead

SPRING

N

E   ✻W

S

SUMMER

below the horizon

FALL

below the horizon

WINTER

N

E   W

✻

S

✻ = CANIS MINOR   ⌐ = BIG DIPPER   • = NORTH STAR   ▪ = OVERHEAD   ⌐ = HORIZON

Maps depict mid-season at 11pm DST/10pm ST (May 5, Aug 5, Nov 5 and Feb 5).
See page 9 for other dates when maps are exact.

CANIS MINOR *(CANE-iss MY-ner)*

English Name: the little dog

Size: very small, 71st largest

When to Look: most prominent from January through March and visible in the late evening sky from December through April

Notes: A small but prominent constellation closely associated with its bright neighbors, Canis Major and Orion. It is defined by just two primary stars. Its main distinction comes from its first-magnitude star Procyon, whose name means "before the dog." Its rise signals the coming of Sirius, the dog star.

WHERE TO LOOK: HORIZON GRAPH

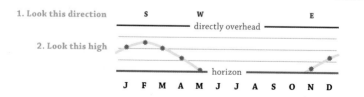

1. Look this direction

S   W   E

directly overhead

2. Look this high

horizon

J F M A M J J A S O N D

Shown 11pm DST/10pm ST on the 15th. If viewing earlier/later, adjust 1 month for every 2 hours. (At 8pm in Jan, use Dec; at 12am in Jan, use Feb.) See page 10 for more info.

Stars in Canis Minor

	LIGHT YEARS	MAGNITUDE	NAME ORIGIN
Procyon (α)	11.4	0.4	Greek for "pre-dog"
Gomeisa (β)	170	2.9	Arabic for "little bleary-eyed one"

Every star is ranked in terms of brightness, or magnitude. The brighter the star, the lower the magnitude. Stars with negative magnitudes are very bright. The sixth magnitude is the limit of human vision; the brightest stars are first magnitude or less.

Procyon (α): The sixth-brightest star in our sky, with a magnitude of 0.4, Procyon marks the northeast corner of the Winter Triangle. The other two corners are marked by Sirius, in Canis Major, and Betelgeuse, in Orion. The star shares its name with the scientific name of the Raccoon genus, the Greek word *Procyon*, which means "before the dog." It gets this name because it rises before Sirius, which is known as the Dog Star.

Gomeisa (β): The only other prominent star in the constellation, its name means "little bleary-eyed one" in Arabic.

The Milky Way passes between Procyon in Canis Minor and Sirius in Canis Major.

Mythology/History:

Although the constellation has a classical origin, and contains one of the brightest stars in the sky, it does not have any mythology specifically associated with it. Even Procyon, the sixth-brightest star visible from northern latitudes, does not have its own mythology. Instead, it is always tied to its neighbor Sirius, in Canis Major, which shines five times as brightly. Procyon was recognized by the ancient civilizations of Egypt and Babylon, which associated it with a dog. The entire constellation, consisting as it does of just two primary stars, appears to have been first distinguished by the Greeks. This makes it a relatively late addition to the ancient constellations.

In Greek mythology, the stories told about Canis Minor are mostly the same as the ones told about Canis Major. In terms of mythology, the two constellations are nearly interchangeable. Occasionally, it is said to represent the dog belonging to Icarius, the first man to learn the secret of wine making from the god Dionysus. The full story of Icarius and his wrongful death is related in the entry for the constellation Boötes (page 51). Today, Canis Minor is often said to be Orion's second dog. In different tellings, it is said to be either another hunting dog or a lap dog. Both of these appear to be modern interpretations, for classical references that mention Orion having a dog only discuss one. Furthermore, there are no references to Orion's dog independent of the stories about the constellations.

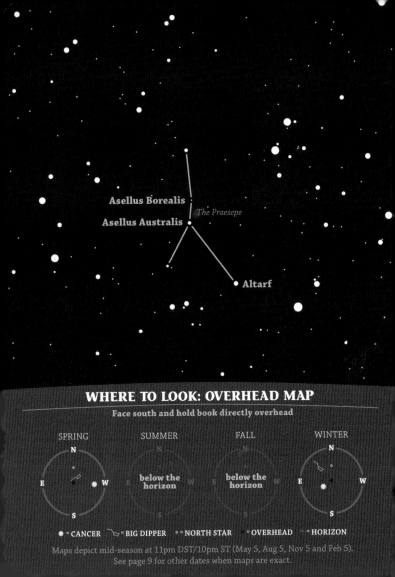

Asellus Borealis

The Praesepe

Asellus Australis

Altarf

WHERE TO LOOK: OVERHEAD MAP

Face south and hold book directly overhead

SPRING

SUMMER

below the horizon

FALL

below the horizon

WINTER

✸ = CANCER   ⌐☌ = BIG DIPPER   • = NORTH STAR   •• = OVERHEAD   ⌐ = HORIZON

Maps depict mid-season at 11pm DST/10pm ST (May 5, Aug 5, Nov 5 and Feb 5).
See page 9 for other dates when maps are exact.

CANCER *(CAN-ser)*

zodiac

English Name: the crab

Size: medium, 31st largest

When to Look: most prominent from January through April and visible in the late evening sky from December through May

Notes: The dimmest constellation in the zodiac, with no star brighter than the fourth magnitude, Cancer was historically very important nonetheless. When the zodiac was created about 2,500 years ago, the sun reached its northernmost point in the sky while in this constellation, marking what we now call the summer solstice. The constellation itself looks nothing like a crab.

WHERE TO LOOK: HORIZON GRAPH

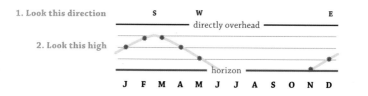

1. Look this direction

S W E

directly overhead

2. Look this high

horizon

J F M A M J J A S O N D

Shown 11pm DST/10pm ST on the 15th. If viewing earlier/later, adjust 1 month for every 2 hours. (At 8pm in Jan, use Dec; at 12am in Jan, use Feb.) See page 10 for more info.

Stars in Cancer

	LIGHT YEARS	MAGNITUDE	NAME ORIGIN
Altarf (β)	290	3.5	Arabic for "the tip"
Asellus Australis (δ)	155	3.9	Latin for "southern donkey"
Asellus Borealis (γ)	155	4.7	Latin for "northern donkey"

Every star is ranked in terms of brightness, or magnitude. The brighter the star, the lower the magnitude. Stars with negative magnitudes are very bright. The sixth magnitude is the limit of human vision; the brightest stars are first magnitude or less.

Altarf (β): The brightest star in the constellation, though still only magnitude 3.5.

Asellus Australis (δ): The constellation's second-brightest star, it lies on the south side of the Praesepe.

Asellus Borealis (γ): Sits across the Praesepe from Asellus Australis, about 4 degrees to the north. Visible on clear, dark nights.

The Praesepe (M44): Visible to the naked eye on clear, dark nights as a faint, fuzzy circle about three times the diameter of the full moon. Its brightest individual stars are of the seventh magnitude and cannot be distinguished with the naked eye. Binoculars reveal a dense bunching of stars that seem to swarm like bees, giving rise to the nickname of "the Beehive Cluster." The Praesepe was one of the first object that Galileo observed with the newly invented telescope in the early 1600s. Galileo's discovery that there were more stars than the eye could see challenged the Ptolemaic model of the heavens and helped pave the way for the scientific revolution of the seventeenth century.

Ecliptic: The sun passes through Cancer from July 19 to August 7. When the zodiac was created, the sun entered Cancer on the summer solstice and its astrological period was set from June 21 to July 21.

Mythology/History:

Although dim, this constellation has been well known since ancient times. About 2,500 years ago, the sun reached its northernmost point in the sky while in Cancer before it began to move backward, like a crab, to the south. At this time of year, which we call the summer solstice, the sun was directly overhead at 23.5 degrees north latitude. To this day, this line is known as the Tropic of Cancer, even though the sun is now in Gemini on the solstice. In Greek mythology, this constellation is said to be the crab that the goddess Hera sent to foil Hercules while he was fighting the Lernean Hydra (page 71). The crab nipped at Hercules' heels to distract him, but the hero stepped on the poor creature, crushing it, and won his battle with the Hydra. Hera placed the crab in the heavens to honor its bravery and loyalty.

The Greeks also have a myth about the stars in the middle of the constellation. Asellus Australis and Asellus Borealis are said to be the donkeys that the gods Dionysus and Silenus rode into battle against the Titans—giants who ruled over the Earth. The donkeys' loud braying unnerved the Titans, allowing the gods of Olympus to triumph. The Praesepe, which means manger in Latin, lies between these two stars and is said to be the donkeys' feeding trough. The astrological symbol for Cancer is (♋) and represents the claws of the crab.

249

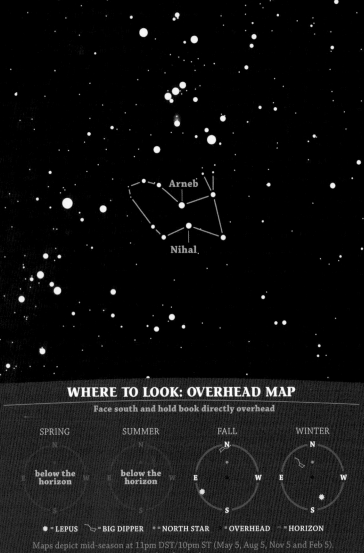

Arneb

Nihal

WHERE TO LOOK: OVERHEAD MAP

Face south and hold book directly overhead

SPRING

below the horizon

SUMMER

below the horizon

FALL

WINTER

✳ = LEPUS 　〰 = BIG DIPPER 　• = NORTH STAR 　• = OVERHEAD 　= HORIZON

Maps depict mid-season at 11pm DST/10pm ST (May 5, Aug 5, Nov 5 and Feb 5).
See page 9 for other dates when maps are exact.

LEPUS (LEP-us)

English Name: the hare

Size: small, 51st largest

When to Look: most prominent in January and visible in the late evening sky from November through February

Notes: This constellation looks quite a bit like its namesake and sits in a rich area of the winter sky. On dark, clear nights it is not difficult to trace the compact shape of a hare running under the feet of the great hunter, Orion. The constellation is easy to locate, but also easy to overlook because of its brilliant neighbors.

WHERE TO LOOK: HORIZON GRAPH

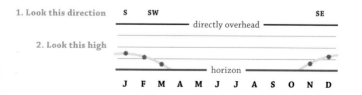

Shown 11pm DST/10pm ST on the 15th. If viewing earlier/later, adjust 1 month for every 2 hours. (At 8pm in Jan, use Dec; at 12am in Jan, use Feb.) See page 10 for more info.

Stars in Lepus

	LIGHT YEARS	MAGNITUDE	NAME ORIGIN
Arneb (α)	1300	2.6	Arabic for "hare"
Nihal (β)	159	2.8	Arabic for the phrase "the camels beginning to quench their thirst"

Every star is ranked in terms of brightness, or magnitude. The brighter the star, the lower the magnitude. Stars with negative magnitudes are very bright. The sixth magnitude is the limit of human vision; the brightest stars are first magnitude or less.

Arneb (α): It is the brightest star in the constellation, with a magnitude of 2.6. Its name comes from the Arabic word for "hare."

Nihal (β): With a magnitude of 2.8, it is the second-brightest star in the constellation. Its name derives from an Arabic phrase meaning "the camels beginning to quench their thirst." The name was originally applied to a small, forgotten Arab constellation consisting of the four brightest stars of Lepus.

Mythology/History:

Though of Greek origin, Lepus has little mythology associated with it. The constellation is usually associated with Orion and is often described as the hunter's quarry, running between his feet. Alternately, the hare is described as being chased by Orion's hunting dog, Canis Major, which follows it across the sky at night. Some ancient authors, however, disputed that a hunter of Orion's stature would spend time chasing a lowly hare and proposed two other sources for this constellation. Some suggested that Hermes (Mercury in Latin), the messenger of the gods, placed the hare in the sky to honor its swiftness.

The other explanation is an interesting moral tale that has its origins in history rather than in mythology. According to legend, there once were no hares on the island of Leros. One day, a boy brought a pregnant hare to the island. When the babies were born, many townspeople took an interest in raising hares. Before long, the island was overrun, as the hares devoured the crops, bringing on a famine. Facing starvation, the people drove the hares from the island through great effort. The image of the hare was placed in the stars as a reminder that too much of a good thing can lead to disaster. This story seems to have a basis in fact, as there is a historical record of a plague of hares on the island of Astypalaea, which is near Leros, around the year 250 BC.

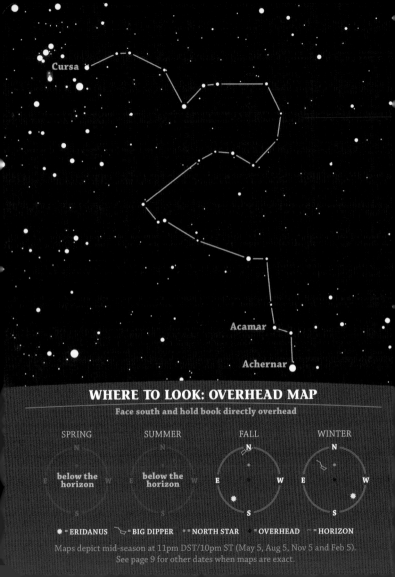

Cursa

Acamar

Achernar

WHERE TO LOOK: OVERHEAD MAP

Face south and hold book directly overhead

SPRING

below the
horizon

SUMMER

below the
horizon

FALL

WINTER

✳ = ERIDANUS   ⬲ = BIG DIPPER   • = NORTH STAR   • = OVERHEAD   = HORIZON

Maps depict mid-season at 11pm DST/10pm ST (May 5, Aug 5, Nov 5 and Feb 5).
See page 9 for other dates when maps are exact.

ERIDANUS *(ih-RID-un-us)*

English Name: the river

Size: very large, 6th largest

When to Look: visible in the late evening sky from November through January

Notes: One of the largest constellations in the sky, though only partially visible from our region. Made up of mostly dimmer stars, and somewhat indistinct in shape, it is difficult to trace and hard to spot unless the skies are quite dark. Its one bright star, the first-magnitude Achernar, is too far south to be visible from our region.

WHERE TO LOOK: HORIZON GRAPH

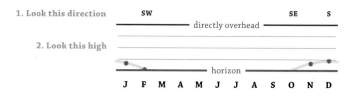

1. Look this direction

SW   SE   S

—————— directly overhead ——————

2. Look this high

——————— horizon ———————

J F M A M J J A S O N D

Shown 11pm DST/10pm ST on the 15th. If viewing earlier/later, adjust 1 month for every 2 hours. (At 8pm in Jan, use Dec; at 12am in Jan, use Feb.) See page 10 for more info.

Stars in Eridanus

	LIGHT YEARS	MAGNITUDE	NAME ORIGIN
Achernar (α)	140	0.5	Arabic for "river's end"
Cursa (β)	85	2.8	Arabic for "footstool of the giant"
Acamar (θ)	160	3.1	Origin unclear

Every star is ranked in terms of brightness, or magnitude. The brighter the star, the lower the magnitude. Stars with negative magnitudes are very bright. The sixth magnitude is the limit of human vision; the brightest stars are first magnitude or less.

Achernar (α): The constellation's brightest star, and one of the brightest stars visible from Earth. Its name is Arabic for "river's end." It is too far south to see except from the southernmost parts of the United States and was not originally part of the constellation cataloged by Ptolemy, but was added later. Because it is so far south, it is not included on this book's list of brightest stars.

Cursa (β): The brightest star far enough north in the constellation to see from most of the United States. The name Cursa comes from an Arabic phrase meaning "footstool of the giant." It marks the northern end of the celestial river.

Acamar (θ): The southern end of the river in classical times; the Greeks considered Acamar the most southern star visible.

Mythology/History:

An ancient constellation, the name Eridanus is Greek in origin. Greek authors disagreed about which river Eridanus represented. Many associated it with the Nile, while others said that it was the Po in what is now Italy, or the Rhone in what is now Switzerland and France. In Roman classical literature, it is universally associated with the Po.

In classical mythology, Eridanus is associated with the myth of Phaethon, the son of Helios, the sun god. According to myth, the young and enthusiastic Phaethon begged his father to let him drive the chariot of the sun across the sky. Reluctantly, Helios agreed, cautioning his son to carefully follow the chariot's well-worn path across the sky. But the youth did not have the strength to control the chariot, and it careened wildly across the sky and fell so low that the Earth caught fire. It is said that this is when the seas dried up, Libya became a desert, and the Ethiopians' skin became dark. To put an end to the chaos, Zeus struck Phaethon down with a thunderbolt. The boy fell, blazing like a falling star, into the river Eridanus. Later, when the Argonauts sailed up the river, they found his body, still smoldering. Eridanus is the tallest of all the constellations, extending from the foot of Orion to 57 degrees south latitude, well below our horizon. In Greek times, the constellation was smaller, ending at the star Acamar.

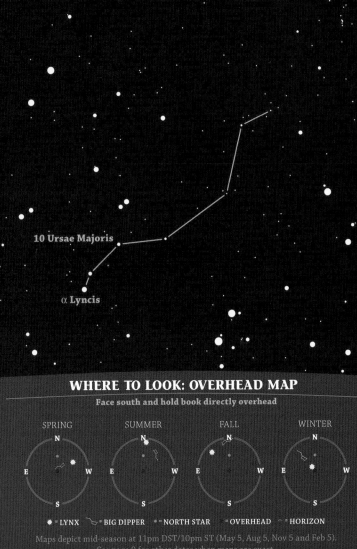

10 Ursae Majoris

α Lyncis

WHERE TO LOOK: OVERHEAD MAP

Face south and hold book directly overhead

SPRING	SUMMER	FALL	WINTER
N	N	N	N
E W	E W	E W	E W
S	S	S	S

✴ = LYNX   ⌐ = BIG DIPPER   • = NORTH STAR   ▪ = OVERHEAD   ⌐ = HORIZON

Maps depict mid-season at 11pm DST/10pm ST (May 5, Aug 5, Nov 5 and Feb 5).
See page 9 for other dates when maps are exact.

LYNX *(LINKS)*

English Name: the lynx

Size: medium, 28th largest

When to Look: most prominent from January through April and visible in the late evening sky from November through June

Notes: A faint, modern constellation located in an unremarkable area of the sky in between Auriga, Gemini and Ursa Major. Difficult to spot and even harder to trace, Hevelius named this constellation Lynx because he felt only those with the eyesight of this nocturnal animal could see it.

WHERE TO LOOK: HORIZON GRAPH

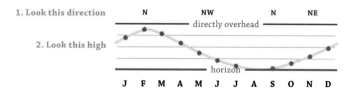

1. Look this direction

2. Look this high

Shown 11pm DST/10pm ST on the 15th. If viewing earlier/later, adjust 1 month for every 2 hours. (At 8pm in Jan, use Dec; at 12am in Jan, use Feb.) See page 10 for more info.

259

Stars in Lynx

	LIGHT YEARS	MAGNITUDE	NAME ORIGIN
α **Lyncis**	220	3.1	No common name
10 Ursae Majoris	46	4.0	See discussion below

Every star is ranked in terms of brightness, or magnitude. The brighter the star, the lower the magnitude. Stars with negative magnitudes are very bright. The sixth magnitude is the limit of human vision; the brightest stars are first magnitude or less.

α **Lyncis**: The brightest star in the constellation, it lies on the constellation's border with the constellation of Leo Minor.

10 Ursae Majoris: A peculiar and somewhat confusing name. When John Flamsteed created his catalog of stars in the early eighteenth century, he considered this star part of Ursa Major. Although it is now placed in Lynx, it has somewhat bizarrely retained its original Flamsteed designation. It is the only star in the sky that carries the name of a different constellation than it belongs to.

Mythology/History:

A modern constellation created by the Polish astronomer Johannes Hevelius in the late seventeenth century to fill a relatively large area of the sky in between the constellations Auriga, Gemini and Ursa Major. Hevelius contributed a total of seven constellations in all to our modern sky charts. (The others are Canes Venatici, Lacerta, Scutum, Leo Minor, Sextans and Vulpecula.)

A dim and indistinct area of the sky, Hevelius says that he named this constellation after the lynx because only observers with eyesight as sharp as the lynx would be able to spot it. It is possible that the name is also a reference to the Greek mythological character Lynceus, one of the Argonauts who sailed with Jason in search of the golden fleece and who was said to have the keenest eyesight in the world.

Hevelius himself must have had the eyes of a lynx, because he was still charting stars with the unaided eye long after most of his colleagues had come to rely on telescopes. The observations he and his wife made for his star catalog were made using a large, non-telescopic astronomical sextant. These observations, some of them made when he was over 70 years old, were considered to be the most accurate of his day, better than most observations made by telescope.

M47

Tureis

Ahadi

Naos

WHERE TO LOOK: OVERHEAD MAP

Face south and hold book directly overhead

SPRING

SUMMER

FALL

WINTER

below the horizon

below the horizon

below the horizon

N
E W
S

✷ = PUPPIS ⌐ = BIG DIPPER • = NORTH STAR ✧ = OVERHEAD ⌢ = HORIZON

Maps depict mid-season at 11pm DST/10pm ST (May 5, Aug 5, Nov 5 and Feb 5).
See page 9 for other dates when maps are exact.

PUPPIS *(PUP-iss)*

English Name: the stern

Size: medium, 20th largest

When to Look: visible in the late evening sky from January through March

Notes: A southern constellation located south and east of the bright stars of Canis Major. A tall constellation, it actually extends farther north than Canis Major, but most of its brighter stars are located far to the south. In the southern end of our range, it is of medium difficulty to spot. Farther north, it becomes increasingly difficult as its brighter stars disappear below the horizon. Its somewhat indistinct shape makes it somewhat difficult to trace, even when all its stars are visible above the horizon.

WHERE TO LOOK: HORIZON GRAPH

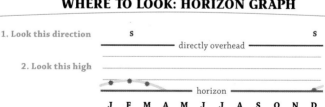

1. Look this direction

2. Look this high

Shown 11pm DST/10pm ST on the 15th. If viewing earlier/later, adjust 1 month for every 2 hours. (At 8pm in Jan, use Dec; at 12am in Jan, use Feb.) See page 10 for more info.

Stars in Puppis

	LIGHT YEARS	MAGNITUDE	NAME ORIGIN
Naos (ζ)	1,100	2.2	Greek for "ship"
Tureis (ρ)	120	2.2	Arabic for "shield"
Ahadi (π)	1,100	2.7	Arabic for "having much promise"

Every star is ranked in terms of brightness, or magnitude. The brighter the star, the lower the magnitude. Stars with negative magnitudes are very bright. The sixth magnitude is the limit of human vision; the brightest stars are first magnitude or less.

Naos (ζ): The brightest star in the constellation, Naos is one of the hottest, most luminous stars in the galaxy with a total power output nearly 800,000 times that of our sun, though much of this is in the ultraviolet spectrum. A short-lived star, it is expected to grow into a red supergiant and end its life in a supernova some time in the next 2 million years. When it explodes, it will appear in the Earth's sky much brighter than the full moon and cast strong shadows at night for months on end.

Tureis (ρ): The easiest to spot of the constellation's brighter stars.

Ahadi (π): The constellation's second-brightest star also lies about 1,100 light years from Earth, yet still shines with an apparent magnitude of 2.7.

M47: An open star cluster in the northern end of the constellation, M47 is visible to the naked eye under ideal conditions as a tiny fuzzy patch. It is easy to spot in binoculars, which resolve several individual stars. The cluster lies about 1,600 light years from Earth, covers an area about the size of the full moon, and has a total apparent magnitude of 4.2.

The Milky Way runs through Puppis, making this a rich constellation to explore with binoculars.

Mythology/History:

The stars of Puppis were part of the constellation Argo Navis, the largest of all the ancient Greek constellations. In 1752, French astronomer Nicolas Louis de Lacaille divided it into three parts: Carina (the keel of the ship), Puppis (the stern of the ship), and Vela (the sails). In Greek mythology, *Argo Navis* was the first ship to set sail and the first Greek naval expedition was Jason and the Argonauts' quest for the golden fleece.

This colorful story featured an all-star cast of Greek heroes and was a favorite subject of both Greek and Roman authors. The golden fleece came from a ram sent by Zeus to save two young children from sacrifice. Phrixus and Helle were the children of King Athamas of Boeotia. Their stepmother, Ino, detested them and plotted their murder. Zeus sent the magical ram to carry the children away. The children climbed on its back and were carried off. Helle lost her grip and fell into the water at the point that separates Europe from Asia. The Greeks called this place the Hellespont in remembrance. Phrixus was delivered safely to the land of Colchis.

Upon arriving, the ram left its fleece to Colchis's king, Æetes, as a gift. Jason and his crew sailed to Colchis, where they captured the golden fleece. Upon their return, their ship was given to Poseidon, the god of the sea, who placed it in the stars.

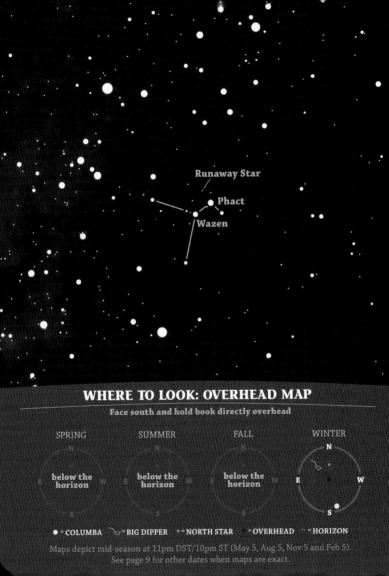

Runaway Star

Phact

Wazen

WHERE TO LOOK: OVERHEAD MAP

Face south and hold book directly overhead

SPRING

SUMMER

FALL

WINTER

below the horizon

below the horizon

below the horizon

* = COLUMBA   = BIG DIPPER   = NORTH STAR   = OVERHEAD   = HORIZON

Maps depict mid-season at 11pm DST/10pm ST (May 5, Aug 5, Nov 5 and Feb 5).
See page 9 for other dates when maps are exact.

COLUMBA (cuh-LUM-buh)

English Name: the dove

Size: medium, 25th largest

When to Look: visible in the late evening sky from January through February

Notes: A small, modern constellation in the southern skies. Though only seen close to the horizon, it is the brightest modern constellation visible from our latitudes and has a compact shape. Under good conditions, it can be spotted and traced with only moderate effort. It does, however, take considerable imagination to see a dove in these stars.

WHERE TO LOOK: HORIZON GRAPH

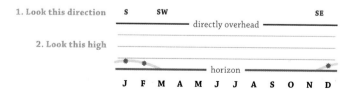

Shown 11pm DST/10pm ST on the 15th. If viewing earlier/later, adjust 1 month for every 2 hours. (At 8pm in Jan, use Dec; at 12am in Jan, use Feb.) See page 10 for more info.

Stars in Columba

	LIGHT YEARS	MAGNITUDE	NAME ORIGIN
Phact (α)	268	2.6	Arabic for "dove"
Wazen (β)	86	3.1	Arabic for "weight"
Runaway Star (μ)	1,300	5.1	Name stems from English

Every star is ranked in terms of brightness, or magnitude. The brighter the star, the lower the magnitude. Stars with negative magnitudes are very bright. The sixth magnitude is the limit of human vision; the brightest stars are first magnitude or less.

Phact (α): This star is not only the brightest in this constellation, but it is the brightest among all the modern constellations visible from our region.

Wazen (β): The second-brightest star in the constellation, with a magnitude of 3.1.

Runaway Star (μ): Just barely visible to the unaided eye under very dark skies, this dim star is one of the fastest moving stars in the galaxy. It is moving relative to the stars around it at a speed of 120 miles per second. Together with AE Aurigae and 53 Arietis (neither of which are bright enough to be seen with the naked eye), this star has a trajectory that traces back to the Trapezium cluster in the Orion Nebula 2.5 million years ago. It is likely that a collision between newly-formed stars in that nebula sent these three hurtling off through the galaxy.

Mythology/History:

Columba first appeared on the star charts published by Dutch astronomer and theologian Petrus Plancius in 1592. Created out of "unformed" stars that were usually considered to be part of Canis Major, Columba, like many of the 20 constellations created by Plancius, has a biblical association. Plancius was actively engaged in a project of "de-paganizing" the heavens by associating constellations with biblical stories. He associated the stars of the *Argo* (the ship sailed by Jason and his crew in search of the golden fleece, pages 195 and 265) with Noah's Ark.

Plancius depicted Columba as the dove that Noah sent out from the ark to seek dry land. The dove returned carrying an olive branch in its beak, a sign that the waters of the Great Flood were receding.

The association of these stars with a dove may have ancient roots, as there are references to a dove in this part of the heavens dating back nearly 2,000 years. The brightest star in the constellation, Phact, appears to get its name from the Arabic word for dove. Scholars speculate that this constellation once represented the dove that Jason and the crew of the *Argo* sent out when they encountered the infamous Clashing Rocks, which crushed any ship that attempted to pass through. The dove flew between and triggered them, allowing the *Argo* to pass through.

269

α Monocerotis

β Monocerotis

WHERE TO LOOK: OVERHEAD MAP

Face south and hold book directly overhead

SPRING	SUMMER	FALL	WINTER
below the horizon	below the horizon		

✳ = MONOCEROS ⌐⌐ = BIG DIPPER • = NORTH STAR ▪ = OVERHEAD ⚬ = HORIZON

Maps depict mid-season at 11pm DST/10pm ST (May 5, Aug 5, Nov 5 and Feb 5).
See page 9 for other dates when maps are exact.

MONOCEROS *(muh-NAH-ser-us)*

English Name: the unicorn

Size: medium, 35th largest

When to Look: most prominent in February and visible in the late evening sky from December through April

Notes: A dim, modern constellation located in the Milky Way. Although it is difficult to see its stars and trace its shape, the region of the sky is very easy to find as it is surrounded by the three first-magnitude stars that form the Winter Triangle: Sirius in Canis Major, Procyon in Canis Minor, and Betelgeuse in Orion.

WHERE TO LOOK: HORIZON GRAPH

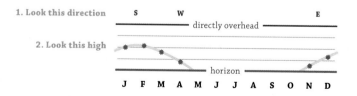

1. Look this direction

S   W   E

— directly overhead —

2. Look this high

— horizon —

J F M A M J J A S O N D

Shown 11pm DST/10pm ST on the 15th. If viewing earlier/later, adjust 1 month for every 2 hours. (At 8pm in Jan, use Dec; at 12am in Jan, use Feb.) See page 10 for more info.

Stars in Monoceros

	LIGHT YEARS	MAGNITUDE	NAME ORIGIN
β **Monocerotis**	715	3.8	No common name
α **Monocerotis**	175	3.9	No common name

Every star is ranked in terms of brightness, or magnitude. The brighter the star, the lower the magnitude. Stars with negative magnitudes are very bright. The sixth magnitude is the limit of human vision; the brightest stars are first magnitude or less.

The stars in this constellation are unspectacular, especially in contrast with their brilliant neighbors in Orion, Canis Major and Canis Minor, which make up the Winter Triangle. None of them have common names and the brightest stars are just of the fourth magnitude. However, the Milky Way passes through the center of this constellation. It is full of faint star clusters and nebulae and is a wonderful area to scan with binoculars.

β **Monocerotis**: The brightest star in the constellation, with a magnitude of 3.8. It is actually a triple star system, though a telescope is needed to resolve its individual parts.

α **Monocerotis**: Slightly dimmer than the constellation's beta star, and situated on the opposite side of the Milky Way.

Mythology/History:

Monoceros is a modern constellation created in 1613 by the Dutch astronomer, cartographer and minister Petrus Plancius. Plancius created 20 constellations, which he published on two star globes.

The unicorn first appeared in European myth in the eighth century BC. It was described in Greek mythology as a creature with the body of a white donkey, a purple head, and a single red, blue and black horn that could cure poisoning.

It is likely that this fantastic creature grew out of accounts of the Indian rhinoceros. The unicorn appears in the Old Testament (in Psalms, Job and Numbers) and by the fourth century AD the creature was being associated with Christ. The unicorn was said to be impossible to hunt and could only be captured by a virgin. By the 1500s, the unicorn was depicted as a magnificent white horse with the beard of a goat and a single white horn. At the same time, there were doubts about its existence; spiral horns that were said to have come from the creatures were the strongest pieces of evidence present. In the mid-1500s an Italian scholar showed that these horns were from narwhals, whales that live in arctic waters. About the same time, the Catholic Church officially denied the existence of the unicorn, citing the mounting evidence that the unicorn was only mythological.

α **Camelopardalis**

β **Camelopardalis**

WHERE TO LOOK: OVERHEAD MAP

Face south and hold book directly overhead

SPRING   SUMMER   FALL   WINTER

✷ = CAMELOPARDALIS   = BIG DIPPER   = NORTH STAR   = OVERHEAD   = HORIZON

Maps depict mid-season at 11pm DST/10pm ST (May 5, Aug 5, Nov 5 and Feb 5).
See page 9 for other dates when maps are exact.

CAMELOPARDALIS *(cuh-MEL-oh-PAR-duh-liss)*

English Name: the giraffe

Size: large, 18th largest

When to Look: most prominent from November through February and visible in the late evening sky throughout the year

Notes: A modern constellation created to fill a large area in the northern sky with no bright stars. Large in area, but very dim, with no stars brighter than the fourth magnitude. It is one of the most difficult constellations to spot and trace. Unlike many constellations, however, it actually looks like its namesake. Under ideal conditions, you really can make out the shape of a giraffe.

WHERE TO LOOK: HORIZON GRAPH

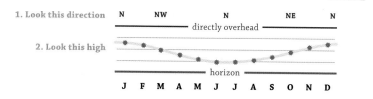

1. Look this direction   N   NW   N   NE   N

directly overhead

2. Look this high

horizon

J F M A M J J A S O N D

Shown 11pm DST/10pm ST on the 15th. If viewing earlier/later, adjust 1 month for every 2 hours. (At 8pm in Jan, use Dec; at 12am in Jan, use Feb.) See page 10 for more info.

Stars in Camelopardalis

	LIGHT YEARS	MAGNITUDE	NAME ORIGIN
β **Camelopardalis**	1,500	4.0	No common name
α **Camelopardalis**	2,800	4.3	No common name

Every star is ranked in terms of brightness, or magnitude. The brighter the star, the lower the magnitude. Stars with negative magnitudes are very bright. The sixth magnitude is the limit of human vision; the brightest stars are first magnitude or less.

β **Camelopardalis**: The constellation's brightest star, but still only magnitude 4.0.

α **Camelopardalis**: The constellation's second-brightest star.

Mythology/History:

A modern constellation created in 1613 by the Dutch astronomer Petrus Plancius. His contributions to astronomy came with the creation of 20 new constellations that he published on two star globes. The first, which was produced in 1598, defined 12 new constellations in the southern skies. These skies had not yet been charted by Europeans, and star charts were needed for navigation. Plancius's second star globe, produced in 1613, defined eight more constellations, mostly in undefined regions of the northern skies.

All of Plancius's first group of constellations are recognized today as modern constellations. Among his second group, only Camelopardalis and Monoceros survived. Most of these constellations had a biblical origin, as Plancius was also a minister. If Plancius had a biblical association in mind for Camelopardalis, it has been lost.

Camelopardalis is a large constellation formed in a region of the sky with no bright stars. The Greeks considered this area empty and called it "the desert." *Camelopardalis* is the Latin word for "giraffe," but the word originated in Greek and literally means "camel leopard," as the Greeks thought that the creature had the head of a camel and the spots of a leopard.

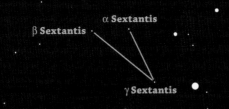

β **Sextantis**   α **Sextantis**

γ **Sextantis**

WHERE TO LOOK: OVERHEAD MAP

Face south and hold book directly overhead

SPRING   SUMMER   FALL   WINTER

below the horizon   **below the horizon**

✳ = SEXTANS   ↘ = BIG DIPPER   • = NORTH STAR   • = OVERHEAD   ⌒ = HORIZON

Maps depict mid-season at 11pm DST/10pm ST (May 5, Aug 5, Nov 5 and Feb 5).
See page 9 for other dates when maps are exact.

SEXTANS *(SEX-tunz)*

English Name: the sextant

Size: small, 47th largest

When to Look: most prominent in March and visible in the late evening sky from January through May

Notes: A very faint modern constellation in an undistinguished area of the sky. Very difficult to spot as it includes no stars brighter than fifth magnitude, it is a good challenge for a stargazer.

WHERE TO LOOK: HORIZON GRAPH

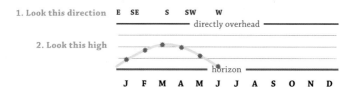

Shown 11pm DST/10pm ST on the 15th. If viewing earlier/later, adjust 1 month for every 2 hours. (At 8pm in Jan, use Dec; at 12am in Jan, use Feb.) See page 10 for more info.

Stars in Sextans

	LIGHT YEARS	MAGNITUDE	NAME ORIGIN
α **Sextantis**	285	4.5	No common name
β **Sextantis**	325	5.1	No common name
γ **Sextantis**	260	5.1	No common name

Every star is ranked in terms of brightness, or magnitude. The brighter the star, the lower the magnitude. Stars with negative magnitudes are very bright. The sixth magnitude is the limit of human vision; the brightest stars are first magnitude or less.

One of the faintest constellations in the entire sky, it contains no stars brighter than fifth magnitude. None of its stars have common names, and it contains no other objects of interest visible to the naked eye.

α **Sextantis**: The brightest star in the constellation, with a magnitude of 4.5.

β **Sextantis**: Magnitude 5.1, distance 520 light years.

γ **Sextantis**: Magnitude 5.1, distance 620 light years.

Mythology/History:

This constellation was created by Polish astronomer Johannes Hevelius in the seventeenth century in honor of his astronomical sextant. A sextant is a device used to precisely measure the positions of celestial objects. The mariner's sextant was a critical tool for open ocean navigation until the latter part of the twentieth century. Navigators used the positions of the sun and stars to determine their ships' latitude and longitude. The astronomical sextant is a much larger and more precise instrument than the mariner's sextant.

With his wife's assistance, Hevelius used his specially-made astronomical sextants to create the most accurate star maps of his time and the first detailed charts of the moon. Although his sextant was not telescopic, the charts that Hevelius and his wife created were as accurate as those made later by Edmond Halley using telescopes.

Hevelius's star maps, which showed the precise locations of 1,564 stars, were published by his wife in 1690 following his death three years earlier. The maps defined several new constellations in areas of the sky that had previously been considered "unformed." Seven of these constellations, including Sextans, are officially recognized today. The other six are Canes Venatici, Lacerta, Leo Minor, Lynx, Scutum and Vulpecula.

γ **Pyxidis**

Alsumut

β **Pyxidis**

WHERE TO LOOK: OVERHEAD MAP

Face south and hold book directly overhead

SPRING	SUMMER	FALL	WINTER
N	N	N	N
below the horizon	**below the horizon**	**below the horizon**	
E · W	E · W	E · W	E · W
S	S	S	S

✴ = PYXIS   ⌐⌐ = BIG DIPPER   ▪ = NORTH STAR   ▪ = OVERHEAD   ▫ = HORIZON

Maps depict mid-season at 11pm DST/10pm ST (May 5, Aug 5, Nov 5 and Feb 5).
See page 9 for other dates when maps are exact.

PYXIS (PIX-iss)

English Name: the ship's compass

Size: small, 65th largest

When to Look: visible in the late evening sky from February through March

Notes: A small, dim, modern constellation. Its brightest stars are just of the fourth magnitude, making it difficult to spot and hard to trace. Under good conditions, it is fully visible throughout the United States, sitting across the Milky Way from the bright stars of Canis Major, but its faint stars are hard to pick out close to the horizon.

WHERE TO LOOK: HORIZON GRAPH

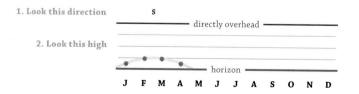

1. Look this direction

S

directly overhead

2. Look this high

horizon

J F M A M J J A S O N D

Shown 11pm DST/10pm ST on the 15th. If viewing earlier/later, adjust 1 month for every 2 hours. (At 8pm in Jan, use Dec; at 12am in Jan, use Feb.) See page 10 for more info.

Stars in Pyxis

	LIGHT YEARS	MAGNITUDE	NAME ORIGIN
Alsumut (α)	845	3.7	Arabic for "the way"
β **Pyxidis**	388	4.0	No common name
γ **Pyxidis**	209	4.0	No common name

Every star is ranked in terms of brightness, or magnitude. The brighter the star, the lower the magnitude. Stars with negative magnitudes are very bright. The sixth magnitude is the limit of human vision; the brightest stars are first magnitude or less.

Alsumut (α): The only named star in the constellation. Its name means "the way" or "the compass bearing" in Arabic, suggesting that these stars were associated with the ship's compass long before de Lacaille created this constellation.

β **Pyxidis**: The constellation's second-brightest star, with a magnitude of 4.0.

γ **Pyxidis**: The dimmest of the three fourth-magnitude stars in the constellation. Though slightly dimmer than β Pyxidis, it also has a magnitude of 4.0.

Mythology/History:

A dim, modern constellation created by French astronomer Nicolas Louis de Lacaille in the mid-eighteenth century, Pyxis represents a mariner's compass. The constellation occupies an area of the sky that was once considered part of the ancient Greek constellation Argo Nevis. Often simply called *Argo*, this was the largest of the ancient constellations and represented Jason and the Argonauts' ship on their quest for the golden fleece.

In the mid-eighteenth century, de Lacaille divided the *Argo* into three constellations: Carina (the keel), Puppis (the stern, page 263), and Vela (the sails). At the same time, he also created the constellation Pyxis Nautica, which is Latin for "nautical box," a term used for a ship's compass.

Although the constellation includes stars that the Greeks associated with the *Argo*, it is not generally considered to be a part of that group of constellations today. When de Lacaille split up the *Argo*, he kept the Bayer designations for its stars. Thus Carina has the α and β; Vela has the γ and δ; and Puppis has the ζ. Pyxis did not contain any of *Argo's* Bayer stars and was given new Bayer star designations, distinguishing it from the three other *Argo* constellations. Today the constellation is known as Pyxis, which simply means "box," but it is still recognized as a ship's compass.

285

α Caelum

Face south and hold book directly overhead

SPRING	SUMMER	FALL	WINTER
below the horizon	below the horizon	below the horizon	

✻ = CAELUM ⌐∿ = BIG DIPPER •= NORTH STAR ⊙ = OVERHEAD = HORIZON

Maps depict mid-season at 11pm DST/10pm ST (May 5, Aug 5, Nov 5 and Feb 5).
See page 9 for other dates when maps are exact.

CAELUM *(SEE-lum)*

English Name: the engraving tool

Size: very small, 81st largest

When to Look: visible in the late evening sky from December through January

Notes: A tiny, dim, modern constellation containing no bright stars. It rises completely above the horizon in the southern half of our range, but its dim stars make it hard to spot so low in the sky.

WHERE TO LOOK: HORIZON GRAPH

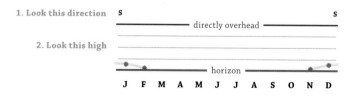

1. Look this direction

2. Look this high

Shown 11pm DST/10pm ST on the 15th. If viewing earlier/later, adjust 1 month for every 2 hours. (At 8pm in Jan, use Dec; at 12am in Jan, use Feb.) See page 10 for more info.

Stars in Caelum

	LIGHT YEARS	MAGNITUDE	NAME ORIGIN
α Caelum	66	4.4	No common name

Every star is ranked in terms of brightness, or magnitude. The brighter the star, the lower the magnitude. Stars with negative magnitudes are very bright. The sixth magnitude is the limit of human vision; the brightest stars are first magnitude or less.

The constellation contains only a single fourth-magnitude star and none of its stars have common names. The constellation does contain a number of very faint galaxies but little of interest for viewers with the unaided eye or binoculars.

α **Caelum**: The brightest star in the constellation, with a magnitude of 4.4. It is a dwarf star and a relatively close neighbor to our sun, just 66 light years away.

Mythology/History:

A modern constellation created by the French astronomer Nicolas Louis de Lacaille in 1756. Lacaille's extensive charts of the southern skies were his most important contributions to astronomy. In 1750, he organized an astronomical expedition to the Cape of Good Hope in what is now South Africa. A shy man who shunned public acclaim, de Lacaille returned to France after his expedition and buried himself in work at Mazarin College. From his observations, de Lacaille produced a catalog of over 10,000 stars and defined 14 new constellations. He also used his detailed observations of the sun and the moon to meticulously calculate a table of eclipses for the following 1,800 years. His work was as high in quality as it was in quantity and he earned universal respect among his peers. He died at the age of 49 from an attack of gout, aggravated by overwork. His astronomical catalog was published posthumously in 1763.

The original name for this constellation was *les Burins*, which is French for "the cold chisel." A cold chisel is a tool used for cutting metals without heating them. It is stronger and blunter than chisels used on wood, and was often used in engraving and print making. Later, de Lacaille gave these stars the Latin name *Caelum Sculptorium*, which means "sculptor's chisel." The International Astronomical Union shortened its official name to Caelum.

STARS, SUPERNOVAE AND GALAXIES

The Life Cycle of a Star

Stars are massive balls of superhot gases. Composed mostly of hydrogen, stars form when gigantic clouds of gas and dust in space are pulled together by the force of gravity, forming a disc-like shape. At the center of this disc, a proto-star forms due to gravity. As the proto-star's gases contract, pressure builds and the material heats up, eventually reaching millions of degrees. At these high pressures and temperatures, hydrogen atoms fuse together to form helium atoms and this nuclear reaction (fusion) releases energy, which lights the star and produces enormous amounts of heat and radiation.

RCW 49 Nebula

LIFE EXPECTANCY

A star's life expectancy depends primarily on its mass. The more massive a star is, the faster it consumes itself. Some stars survive a few million years, while others last for tens of billions of years. The most massive stars tend to burn more brightly and hotter and use their fuel more quickly. Smaller stars tend to burn more slowly and survive longer.

RED GIANTS AND DWARF STARS

When a small- or medium-sized star begins to exhaust all of the hydrogen in its core, it becomes a red giant, growing many times its previous size. As the star depletes its supply of hydrogen, it begins to fuse helium, which gives off more heat. The star expands and becomes a red giant as it radiates off the additional heat. The leftover core inside becomes very dense; this dense core is called a dwarf star and it is so dense that a spoonful would weigh more than a car. The sun will likely meet its end in this way.

SUPERGIANTS

When a large star uses up its hydrogen fuel, it continues its nuclear reactions by fusing heavier and heavier atoms, generating elements like iron in the process. It swells in size, becoming a supergiant. This massive star may become unstable late in life, varying its light output or ejecting large amounts of superhot gases into space.

SUPERNOVAE

Some supergiants reach a point where the fusion reactions in the inner layers of the star can no longer support the outer layers. This causes the outer layers to collapse inward, condensing the core into a very, dense object (a neutron star). This causes an implosion, which then rebounds, creating the explosion we refer to as a supernova. Supernovae cast gas and dust deep into the universe. In fact, ancient supernovae are the sources of the heaviest elements in the universe, including the carbon, oxygen and nitrogen that make up our bodies, as well as the silicon, sodium, aluminum and iron that form the Earth's crust. After a supernova, if the remnants of the star's core are large enough, it will collapse back upon itself, becoming a black hole—a concentration of matter so dense that not even light can escape its gravitational pull.

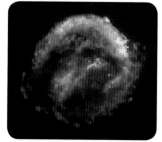

A supernova remnant

Galaxies

A galaxy is an enormous collection of stars, gas and dust. The universe consists of billions of galaxies, which come in many shapes and sizes and can contain anywhere from a few million to over a trillion stars. Our own galaxy, the Milky Way, is a relatively large spiral galaxy made up

The spiral galaxy NGC 4414

of about 200 billion stars. Three other galaxies are visible with the unaided eye, but only the Andromeda Galaxy (M31, page 173)—a spiral galaxy similar to our own—can be seen from the latitudes of the northern hemisphere. At 2.5 million light years away, it is the most distant object visible to the naked eye.

THE SOLAR SYSTEM

The solar system includes the sun and all the objects that orbit it, including the eight planets and their moons, dwarf planets, the asteroid belt, comets, dust and debris.

The Sun

The sun is the largest object in our solar system, making up the vast majority of our solar system's mass—about 99.8 percent. Its mass is the equivalent of about 332,000 Earths.

It turns out that small, dim stars are quite common, making our sun

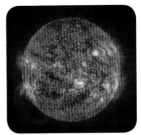
The sun

larger and brighter than 95 percent of all the stars in the Milky Way. Compared to other stars visible with the naked eye, or even with a small telescope, our sun is on the small side of average. Most of the stars we see at night are the giants of the galaxy. Some stars are just one-tenth the mass of our sun, whereas others are a hundred times its mass. In terms of brightness, some stars are one-millionth as bright, while others are a million times brighter.

Sometimes referred to as Sol (Latin for "sun" and root of the word "solar"), the sun is about 93 million miles away from Earth. It takes light eight minutes to travel this distance. The temperature of the sun varies. The sun's surface is about 10,000°F, but its core is much hotter, at about 27 million°F.

The Terrestrial Planets and the Gas Giants

The eight planets

The eight planets of the solar system are often broken into two groups: the terrestrial planets (Mercury, Venus, Earth and Mars) and the gas giants (Jupiter, Saturn, Uranus and Neptune). Terrestrial planets are rocky and contain heavier elements like iron and nickel, while gas giants consist mostly of gases like hydrogen, helium and methane. Gas giants are also much larger than terrestrial planets. Nevertheless, because the components of gas giants are often quite light, they tend to be less dense than the terrestrial planets. For instance, Jupiter is about one-fourth as dense as Earth or about as dense as water. Some gas giants are even less dense. Saturn would float on water if there were an ocean large enough to hold it.

Mercury

Mercury

Orbiting at an average distance of 36 million miles, Mercury is the closest planet to the sun. With a diameter of only about 3,000 miles, it is also our solar system's smallest planet. Mercury has no moon, almost no atmosphere, and the planet's surface is pockmarked with craters from meteorite impacts. In terms of composition and appearance, Mercury's surface is quite similar to that of Earth's moon. Perhaps surprisingly, despite its proximity to the sun, Mercury isn't always an inferno. During the day, temperatures can exceed 800°F. However, at night, the temperature drops below –275°F.

Mercury was named after the Roman messenger god of the same name—a god who traveled quickly through the air. In this respect, the planet was aptly named: Mercury completes a full orbit around the sun faster than any other planet; it takes just 88 Earth days. Nevertheless, Mercury rotates on its axis quite slowly, only once every 59 Earth days. Because of this combination of quick orbit and slow rotation, it takes 176 Earth days for an area to experience a solar day, compared to the twenty-four hours it takes on Earth.

Venus

Venus is sometimes referred to as Earth's "twin," which isn't far off when you consider the size of the two planets: Earth's diameter is only a few hundred miles greater than that of Venus. However, most of the similarities end there. Venus has no moon and is much closer to the sun—67.2 million miles away, compared to Earth's 93 million.

Venus is wrapped in a shroud of heavy clouds and has the thickest atmosphere of any terrestrial planet. The atmospheric pressure on its surface is about 90 times higher than Earth's. Furthermore, while Earth's atmosphere consists primarily of nitrogen and oxygen, Venus's is mostly composed of carbon dioxide, with traces of water vapor and nitrogen.

Venus

Scientists speculate that Venus might once have been quite temperate and perhaps even habitable, with oceans of liquid water. That is certainly not the case today. At 870°F—a temperature high enough to melt lead—Venus has the highest surface temperature in the solar system. However, this high temperature isn't simply the result of the planet's relative proximity to the sun. Instead, its conditions are likely the result of a "runaway greenhouse effect," which occurred because Venus's thick atmosphere trapped heat, producing more greenhouse gases. In turn, these gases trapped even more heat.

Mars

The fourth planet in the solar system, Mars is about 141 million miles away from the sun. Because of this great distance, Mars receives much less sunlight than the Earth and is generally much colder. Temperatures on Mars range from below −200°F in the upper atmosphere to around −20° to −40°F closer to the surface. By comparison, the lowest temperature recorded on Earth was

Mars

−128.6°F, which occurred at Vostok Station in Antarctica in 1983. A Martian year isn't only cold, it's also long. It takes Mars 687 Earth days to orbit the sun.

Named for the Roman god of war because of its blood-red color, Mars gets its hue from the iron-rich minerals on its surface. While mostly made up of wide, flat plains, Mars also has some of the most interesting surface features in the solar system, including Valles Marineris (Latin for *"Valley of Mariner"*), a series of valleys and canyons that stretches about 2,500 miles and reaches depths of up to six miles. It was discovered by the Mariner 9 space probe in the early 1970s. By way of comparison, the Grand Canyon is about 280 miles long and just over a mile deep at its deepest point.

Some of the surface features on Mars also reach great heights. The volcano *Olympus Mons* (Latin for Mount Olympus) is about 17 miles tall and about 370 miles in diameter. Earth's highest mountain, Mount Everest, is only about 29,000 feet tall, or 5.5 miles tall.

Mars is the only terrestrial planet aside from Earth with moons. It has two small, rocky moons named Phobos and Deimos.

Mars is one of the most likely candidates for life in the solar system and is one of the most heavily studied planets. While no life has been conclusively detected, some scientists claim to have evidence of microbial life on the planet. Robotic explorers like the Mars Exploration

Olympus Mons, the largest volcano on Mars

Rovers Spirit and Opportunity have amassed conclusive proof that liquid water—an essential ingredient for life—was once quite common on Mars. This water is no longer present on the surface, but Mars appears to house a significant amount of frozen water beneath its surface, especially at its poles.

The Asteroid Belt

Situated between the orbits of Mars and Jupiter, the asteroid belt is likely the result of collisions between a group of larger rocky bodies, leading to their eventual fragmentation and disintegration. Today, the asteroid belt includes more than 750,000 asteroids with a diameter greater than three-fifths of a mile. There may be more than 1 million. Most of them, however, are relatively small; only 200 or so have a diameter greater than 60 miles. These pale in comparison to the asteroid belt's largest asteroid, Ceres, with a diameter of around 600 miles.

Jupiter

Named for the Roman king of the gods, Jupiter is the largest planet and the most massive body in the solar system other than the sun. Jupiter also has more moons than any other

Jupiter

planet, with 63. Although Jupiter's diameter is 11 times that of the Earth, it is only one-tenth of the sun's.

When Jupiter first formed, a massive amount of material condensed together, producing so much heat that Jupiter emitted more than it received from the sun. This trend continues today, as Jupiter emits about twice the amount of heat it receives.

Jupiter is about 180 million miles from the sun. Because of this distance, it receives much less light and heat than the terrestrial planets, and many areas on Jupiter are quite cold. The temperature at the upper reaches of the atmosphere is about –240°F. However, toward the center of the planet, the temperature reaches an amazing 40,000°F, hotter than the surface of the sun.

Because its atmosphere consists mostly of gases like hydrogen and helium, Jupiter probably doesn't have a solid surface. Nevertheless, as photographs from the Voyager space probe and other spacecraft attest, Jupiter's atmosphere is anything but boring. Photographs of the planet show colorful layers of cloud cover, as well as evidence of high winds and a good deal of atmospheric activity. The best example of this is Jupiter's famous Great Red Spot, a hurricane-like storm that, at its largest, is about three times the size of Earth.

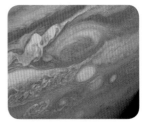

Jupiter's Great Red Spot

Saturn

The sixth planet from the sun, Saturn is composed mostly of hydrogen and helium. Its atmosphere reaches its coldest temperatures (about −280°F) closest to its cloud tops and

Saturn

becomes progressively warmer towards its center and is about 20,000°F at its core. Saturn is smaller than Jupiter but not by much. Jupiter has a diameter of about 88,000 miles, whereas Saturn's diameter is 75,000, about 10 times that of Earth.

Saturn has seven sets of rings, which consist of ice particles ranging in size from tiny flakes and pebbles to boulder-sized chunks. There are competing theories about how the rings formed; some suggest that debris left behind by the formation of our solar system were captured by Saturn's gravity, whereas others contend that the rings are the remains of a natural satellite (moon) destroyed either by Saturn's gravity or by a meteorite impact. While Saturn is distinguished by its rings, it is worth mentioning that Saturn isn't the only planet with them; Jupiter, Uranus and Neptune also have rings—albeit much less conspicuous and famous.

Saturn has 25 moons that are at least six miles in diameter. One of them, Titan, is larger than Mercury and the dwarf-planet Pluto. Titan is particularly interesting because it is one of the few moons to have an atmosphere, as well as liquid hydrocarbon lakes. It is considered a possible candidate for life.

Uranus

The seventh planet from the sun, Uranus is a distant, cold world. On average, Uranus is about 1.7 billion miles away from the sun and receives much less sunlight and heat than the terrestrial planets. Temperatures can fall as low as −355°F. Because of its generally cold temperature, Uranus is sometimes classified as an "ice giant." Nevertheless, the temperature toward the center of the planet is much warmer, perhaps in excess of 4200°F.

Like the rest of the outer planets, Uranus is cloaked in heavy cloud cover. Its atmosphere consists primarily of hydrogen and helium and is interspersed with methane ice crystals, giving the planet a hazy blue-green color. With a diameter four times that of Earth, Uranus is immense, but it is markedly smaller than Saturn and Jupiter, both of which have diameters more than twice that of Uranus. Like its larger neighbors, Uranus has many satellites with at least 21 moons.

Uranus

Uranus is perhaps most notable for its strange rotation upon its axis. It is tilted at 98 degrees relative to the plane of its orbit. In comparison, Earth rotates on its axis at an angle of 23.5 degrees. This means that Uranus is essentially a planet on its side, which creates strange weather patterns.

Neptune

At an average distance of 2.7 billion miles away, Neptune is the farthest planet from the sun and—not surprisingly—the coldest. However, if you were looking for the coldest place in the

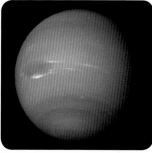

Neptune

solar system, Neptune's largest moon, Triton, has the planet beat. Its surface temperature falls to −390°F. (The coldest temperature possible is called absolute zero or −459.67°F.) For this reason Neptune, along with its neighbor Uranus, is classified as an "ice giant."

Neptune features some of the most volatile weather in the solar system, though the mechanism driving its weather systems is unknown. Nevertheless, Neptune's atmosphere, which consists largely of hydrogen and helium and gets its characteristic blue color from traces of methane, features violent storms and bands of fast-moving clouds. At times, winds here have reached 700 miles per hour. In comparison, the fastest wind speed recorded on Earth was 231 miles per hour at Mount Washington Observatory in New Hampshire; the speed of sound is about 768 miles per hour at sea level on Earth.

Neptune orbits the sun once every 165 Earth years. However, it revolves on its axis every 16 hours and 7 minutes. This means that a year on Neptune would take about 89,000 of its "days."

Neptune is too far away from Earth to see without a telescope. For this reason, the planet went undiscovered until the 1840s, fairly recent by historical standards.

Dwarf Planets and Plutoids

For much of the twentieth century, the solar system boasted nine planets—in part because the term "planet" was never clearly defined. This wasn't problematic until technological advances led to the discovery of many objects at the fringe of the solar system that could be classified as planets. To avoid a sudden influx of new "planets," a clearer definition of the term became necessary. For this reason, the International Astronomical Union voted in 2006 to re-classify Pluto as "a dwarf planet," or an object that meets three criteria: it orbits the sun, it has sufficient mass to assume a round shape, and its gravity has not cleared the area around its own orbit. Put another way, to qualify as a planet an object must be large enough to "clear the neighborhood"; since Pluto's orbit is littered with other objects, it doesn't qualify. If the IAU had not required this last condition, many other objects would likely have to be considered planets, leading to dozens of potential new additions. Nevertheless, this definition is not completely settled and may change in the future.

Of course, Pluto is the most famous dwarf planet, but there are others and some (like Eris) are larger than Pluto. Dwarf planets that are beyond the orbit of Neptune are referred to as Plutoids, in honor of the erstwhile planet. Not all dwarf planets are Plutoids, however. One dwarf planet, Ceres, is closer to the sun; it has a diameter of just under 600 miles and is located in the asteroid belt.

The solar system with the famous dwarf planet, Pluto, in the lower right hand corner.

PLANET LOCATION TABLES AND A LIST OF THE BRIGHTEST STARS

We've included tables to help you find Mercury, Venus, Mars, Jupiter and Saturn. Of the eight planets, these are the easiest to see. Uranus can be visible, but is quite hard to spot. Neptune and Pluto are too distant to be seen with the naked eye. We've also included a list of the brightest stars in the sky; if you familiarize yourself with the brightest stars in the sky, you're well on your way to learning the constellations.

Mercury and Venus

When locating Mercury and Venus, stargazers often refer to whether the planet is a "morning star" or an "evening star." If Mercury or Venus is a "morning star," it will be in the sky *prior to sunrise*. When either planet is an "evening star," it is visible *after sunset*. To locate Mercury when it's a morning star, look in the general vicinity of sunrise. When it's an evening star, look toward sunset. Mercury can be difficult to spot; be sure to look at one of the "best times" listed. Mercury may be visible at the other times, but it is easily lost in the sun's glare. Venus is much easier to see, as it is the brightest object in the sky other than the sun and the moon. Venus is visible as either a morning star or an evening star and is very easy to locate. Just look along the general path of the sun, toward the east in the morning or toward the west in the evening. It will be very easy to spot— Venus is unmistakable.

Mars, Jupiter and Saturn

Mars, Jupiter and Saturn appear to move through the night sky on the line of the ecliptic, which passes through all the constellations of the zodiac, plus Ophiuchus; we've listed the constellation where the planet will be located for a given period of time. Simply find the current date, look up the constellation listed, and look for a bright "star" in the sky that isn't shown on the constellation's graphic.

Mercury

May Be Visible as Morning Star	Best Date in Morning Sky	May Be Visible ss Evening Star	Best Date in Evening Sky
Sep 30 – Oct 14, 2009	Oct 6, 2009	Dec 3 – Dec 28, 2009	Dec 19, 2009
Jan 12 – Feb 23, 2010	Jan 27, 2010	Mar 31 – Apr 18, 2010	Apr 9, 2010
May 9 – Jun 15, 2010	May 26, 2010	Jul 13 – Aug 26, 2010	Aug 7, 2010
Sep 14 – Sep 27, 2010	Sep 20, 2010	Nov 12 – Dec 12, 2010	Dec 2, 2010
Dec 28, 2010 – Feb 3, 2011	Jan 10, 2011	Mar 15 – Mar 30, 2011	Mar 23, 2011
Apr 20 – May 30, 2011	May 8, 2011	Jun 27 – Aug 8, 2011	Jul 20, 2011
Aug 28 – Sep 11, 2011	Sep 3, 2011	Oct 22 – Nov 27, 2011	Nov 14, 2011
Dec 12, 2011 – Jan 13, 2012	Dec 23, 2011	Feb 27 – Mar 12, 2012	Mar 5, 2012
Mar 31 – May 14, 2012	Apr 19, 2012	Jun 10 – Jul 19, 2012	Jul 1, 2012
Aug 9 – Aug 25, 2012	Aug 17, 2012	Oct 1 – Nov 10, 2012	Oct 27, 2012
Nov 26 – Dec 22, 2012	Dec 5, 2012	Feb 9 – Feb 23, 2013	Feb 17, 2013
Mar 13 – Apr 27, 2013	Apr 1, 2013	May 25 – Jun 29, 2013	Jun 13, 2013
Jul 21 – Aug 9, 2013	Jul 30, 2013	Sep 12 – Oct 25, 2013	Oct 9, 2013
Nov 10 – Dec 1, 2013	Nov 18, 2013	Jan 23 – Feb 7, 2014	Jan 31, 2014
Feb 24 – Apr 11, 2014	Mar 14, 2014	May 10 – Jun 9, 2014	May 25, 2014
Jul 1 – Jul 25, 2014	Jul 13, 2014	Aug 25 – Oct 9, 2014	Sep 22, 2014
Oct 26 – Nov 12, 2014	Nov 2, 2014	Jan 4 – Jan 22, 2015	Jan 15, 2015
Feb 7 – Mar 25, 2015	Feb 25, 2015	Apr 25 – May 19, 2015	May 7, 2015
Jun 11 – Jul 10, 2015	Jun 25, 2015	Aug 8 – Sep 23, 2015	Sep 4, 2015
Oct 10 – Oct 24, 2015	Oct 16, 2015	Dec 16, 2015 – Jan 6, 2016	Dec 29, 2015
Jan 22 – Mar 6, 2016	Feb 7, 2016	Apr 8 – Apr 28, 2016	Apr 19, 2016
May 20 – Jun 23, 2016	Jun 5, 2016	Jul 22 – Sep 5, 2016	Aug 17, 2016
Sep 23 – Oct 6, 2016	Sep 29, 2016	Nov 24 – Dec 21, 2016	Dec 11, 2016
Jan 5 – Feb 14, 2017	Jan 19, 2017	Mar 24 – Apr 9, 2017	Apr 1, 2017
Apr 30 – Jun 8, 2017	May 18, 2017	Jul 5 – Aug 18, 2017	Jul 30, 2017
Sep 6 – Sep 20, 2017	Sep 12, 2017	Nov 3 – Dec 5, 2017	Nov 24, 2017
Dec 21, 2017 – Jan 25, 2018	Jan 2, 2018	Mar 8 – Mar 22, 2018	Mar 16, 2018
Apr 11 – May 23, 2018	Apr 30, 2018	Jun 20 – Jul 31, 2018	Jul 12, 2018
Aug 20 – Sep 4, 2018	Aug 27, 2018	Oct 14 – Nov 20, 2018	Nov 7, 2018
Dec 5, 2018 – Jan 4, 2019	Dec 15, 2018	Feb 19 – Mar 5, 2019	Feb 27, 2019
Mar 24 – May 7, 2019	Apr 12, 2019	Jun 4 – Jul 11, 2019	Jun 24, 2019
Aug 1 – Aug 19, 2019	Aug 10, 2019	Sep 24 – Nov 4, 2019	Oct 20, 2019
Nov 20 – Dec 13, 2019	Nov 28, 2019	Feb 2 – Feb 17, 2020	Feb 11, 2020

Venus as a Morning Star

Visible as Morning Star	Highest in Morning Sky	Not Visible
Apr 6 – Nov 7, 2009	Jun 6, 2009	Nov 8, 2009 – Mar 17, 2010
Nov 8, 2010 – Jun 18, 2011	Jan 9, 2011	Jun 19 – Oct 14, 2011
Jun 16, 2012 – Jan 22, 2013	Aug 15, 2012	Jan 23 – May 28, 2013
Jan 20 – Aug 25, 2014	Mar 23, 2014	Aug 26 – Dec 28, 2014
Aug 24, 2015 – Apr 7, 2016	Oct 26, 2015	Apr 8 – Aug 3, 2016
Apr 3 – Nov 4, 2017	Jun 4, 2017	Nov 5, 2017 – Mar 15, 2018
Nov 5, 2018 – Jun 16, 2019	Jan 6, 2019	Jun 17 – Oct 12, 2019
Jun 14, 2020 – Jan 19, 2021	Aug 13, 2020	Jan 20 – May 25, 2021

Venus as an Evening Star

Visible as Evening Star	Highest in Evening Sky	Not Visible
Mar 18 – Oct 19, 2010	Aug 20, 2010	Oct 20 – Nov 7, 2010
Oct 15, 2011 – May 26, 2012	Mar 27, 2012	May 27 – Jun 15, 2012
May 29, 2013 – Jan 1, 2014	Nov 1, 2013	Jan 2 – Jan 19, 2014
Dec 29, 2014 – Aug 6, 2015	Jun 7, 2015	Aug 7 – Aug 23, 2015
Aug 4, 2016 – Mar 16, 2017	Jan 13, 2017	Mar 17 – Apr 2, 2017
Mar 16 – Oct 17, 2018	Aug 18, 2018	Oct 18 – Nov 4, 2018
Oct 13, 2019 – May 24, 2020	Mar 25, 2020	May 25 – Jun 13, 2020
May 26 – Dec 29, 2021	Oct 30, 2021	Dec 30, 2021 – Jan 18, 2022

Mars

Date	Constellation	Date	Constellation
Oct 10 – Nov 24, 2009	Cancer (pg. 247)	Dec 1, 2010 – Jan 13, 2011	Sagittarius (pg. 111)
Nov 25, 2009 – Jan 15, 2010	Leo (pg. 45)	Jan 14 – Feb 17, 2011	Capricornus (pg. 131)
Jan 16 – May 8, 2010	Cancer (pg. 247)	Feb 18 – Mar 19, 2011	Aquarius (pg. 197)
May 9 – Jul 16, 2010	Leo (pg. 45)	Mar 20 – May 2, 2011	Pisces (pg. 201)
Jul 17 – Sep 24, 2010	Virgo (pg. 53)	May 3 – Jun 9, 2011	Aries (pg. 193)
Sep 25 – Oct 27, 2010	Libra (pg. 65)	Jun 10 – Aug 1, 2011	Taurus (pg. 231)
Oct 28 – Nov 5, 2010	Scorpius (pg. 107)	Aug 2 – Sep 12, 2011	Gemini (pg. 235)
Nov 6 – Nov 30, 2010	Ophiucus (pg. 123)	Sep 13 – Oct 16, 2011	Cancer (pg. 247)

Date	Constellation	Date	Constellation
Oct 17, 2011 – Jan 4, 2012	Leo (pg. 45)	Jan 15 – Mar 4, 2016	Libra (pg. 65)
Jan 5 – Feb 12, 2012	Virgo (pg. 53)	Mar 5 – Mar 25, 2016	Scorpius (pg. 107)
Feb 13 – Jun 17, 2012	Leo (pg. 45)	Mar 26 – May 8, 2016	Ophiucus (pg. 123)
Jun 18 – Sep 2, 2012	Virgo (pg. 53)	May 9 – May 27, 2016	Scorpius (pg. 107)
Sep 3 – Oct 6, 2012	Libra (pg. 65)	May 28 – Aug 3, 2016	Libra (pg. 65)
Oct 7 – Oct 15, 2012	Scorpius (pg. 107)	Aug 4 – Aug 18, 2016	Scorpius (pg. 107)
Oct 16 – Nov 10, 2012	Ophiucus (pg. 123)	Aug 19 – Sep 19, 2016	Ophiucus (pg. 123)
Nov 11 – Dec 23, 2012	Sagittarius (pg. 111)	Sep 20 – Nov 6, 2016	Sagittarius (pg. 111)
Dec 24, 2012 – Jan 27, 2013	Capricornus (pg. 131)	Nov 7 – Dec 13, 2016	Capricornus (pg. 131)
Jan 28 – Feb 27, 2013	Aquarius (pg. 197)	Dec 14, 2016 – Jan 14, 2017	Aquarius (pg. 197)
Feb 28 – Apr 16, 2013	Pisces (pg. 201)	Jan 15 – Mar 6, 2017	Pisces (pg. 201)
Apr 17 – May 20, 2013	Aries (pg. 193)	Mar 7 – Apr 9, 2017	Aries (pg. 193)
May 21 – Jul 11, 2013	Taurus (pg. 231)	Apr 10 – Jun 2, 2017	Taurus (pg. 231)
Jul 12 – Aug 22, 2013	Gemini (pg. 235)	Jun 3 – Jul 14, 2017	Gemini (pg. 235)
Aug 23 – Sep 22, 2013	Cancer (pg. 247)	Jul 15 – Aug 14, 2017	Cancer (pg. 247)
Sep 23 – Nov 21, 2013	Leo (pg. 45)	Aug 15 – Oct 9, 2017	Leo (pg. 45)
Nov 22, 2013 – Aug 7, 2014	Virgo (pg. 53)	Oct 10 – Dec 18, 2017	Virgo (pg. 53)
Aug 8 – Sep 13, 2014	Libra (pg. 65)	Dec 19, 2017 – Jan 25, 2018	Libra (pg. 65)
Sep 14 – Sep 23, 2014	Scorpius (pg. 107)	Jan 26 – Feb 5, 2018	Scorpius (pg. 107)
Sep 24 – Oct 19, 2014	Ophiucus (pg. 123)	Feb 6 – Mar 8, 2018	Ophiucus (pg. 123)
Oct 20 – Dec 2, 2014	Sagittarius (pg. 111)	Mar 9 – May 10, 2018	Sagittarius (pg. 111)
Dec 3, 2014 – Jan 6, 2015	Capricornus (pg. 131)	May 11 – Nov 8, 2018	Capricornus (pg. 131)
Jan 7 – Feb 6, 2015	Aquarius (pg. 197)	Nov 9 – Dec 16, 2018	Aquarius (pg. 197)
Feb 7 – Mar 27, 2015	Pisces (pg. 201)	Dec 17, 2018 – Feb 10, 2019	Pisces (pg. 201)
Mar 28 – Mar 31, 2015	Aries (pg. 193)	Feb 11 – Mar 19, 2019	Aries (pg. 193)
April 1 – Jun 22, 2015	Taurus (pg. 231)	Mar 20 – May 14, 2019	Taurus (pg. 231)
Jun 23 – Aug 3, 2015	Gemini (pg. 235)	May 15 – Jun 26, 2019	Gemini (pg. 235)
Aug 4 – Sep 3, 2015	Cancer (pg. 247)	Jun 27 – Jul 27, 2019	Cancer (pg. 247)
Sep 4 – Oct 29, 2015	Leo (pg. 45)	Jul 28 – Sep 21, 2019	Leo (pg. 45)
Oct 30, 2015 – Jan 14, 2016	Virgo (pg. 53)	Sep 22 – Nov 28, 2019	Virgo (pg. 53)

Jupiter

Date	Constellation
Jul 17 – Dec 27, 2009	Capricornus (pg. 131)
Dec 28, 2009 – Apr 11, 2010	Aquarius (pg. 197)
Apr 12, 2010 – May 20, 2011	Pisces (pg. 201)
May 21, 2011 – May 7, 2012	Aries (pg. 193)
May 8, 2012 – Jun 19, 2013	Taurus (pg. 231)
Jun 20, 2013 – Jun 29, 2014	Gemini (pg. 235)
Jun 30 – Oct 3, 2014	Cancer (pg. 247)
Oct 4, 2014 – Feb 17, 2015	Leo (pg. 45)
Feb 18 – May 29, 2015	Cancer (pg. 247)
May 30 – Dec 9, 2015	Leo (pg. 45)
Dec 10, 2015 – Feb 8, 2016	Virgo (pg. 53)
Feb 9 – Jul 30, 2016	Leo (pg. 45)
Jul 31, 2016 – Nov 6, 2017	Virgo (pg. 53)
Nov 7, 2017 – Nov 5, 2018	Libra (pg. 65)
Nov 6 – Dec 5, 2018	Scorpius (pg. 107)
Dec 6, 2018 – Nov 8, 2019	Ophiucus (pg. 123)

Saturn

Date	Constellation
Aug 20, 2009 – Nov 20, 2012	Virgo (pg. 53)
Nov 21, 2012 – Jun 12, 2013	Libra (pg. 65)
Jun 13 – Aug 4, 2013	Virgo (pg. 53)
Aug 5, 2013 – Dec 15, 2014	Libra (pg. 65)
Dec 16, 2014 – Jul 1, 2015	Scorpius (pg. 107)
Jul 2 – Sep 3, 2015	Libra (pg. 65)
Sep 4 – Nov 16, 2015	Scorpius (pg. 107)
Nov 17, 2015 – Feb 1, 2017	Ophiucus (pg. 123)
Feb 2 – Jun 13, 2017	Sagittarius (pg. 111)
Jun 14 – Nov 1, 2017	Ophiucus (pg. 123)
Nov 20, 2017 – Mar 1, 2020	Sagittarius (pg. 111)

A List of the Brightest Stars

Rank	Magnitude	Proper Name	Constellation	Distance (ly)
	−26.73	Sun (Sol)		0.000016
1	−1.47	Sirius	Canis Major (pg. 227)	8.6
2	−0.04	Arcturus	Boötes (pg. 49)	37
3	0.03	Vega	Lyra (pg. 99)	25
4	0.07	Capella	Auriga (pg. 239)	42
5	0.112	Rigel	Orion (pg. 223)	770
6	0.34	Procyon	Canis Minor (pg. 243)	11
7	0.58	Betelgeuse	Orion (pg. 223)	640
8	0.77	Altair	Aquila (pg. 103)	17
9	0.85	Aldebaran	Taurus (pg. 231)	65
10	1.04	Spica	Virgo (pg. 53)	260
11	1.09	Antares	Scorpius (pg. 107)	600
12	1.15	Pollux	Gemini (pg. 235)	34
13	1.16	Fomalhaut	Piscis Austrinus (pg. 189)	25
14	1.25	Deneb	Cygnus (pg. 95)	1550
15	1.35	Regulus	Leo (pg. 45)	77
16	1.5	Alnitak	Orion (pg. 223)	820
17	1.51	Adhara	Canis Major (pg. 227)	430
18	1.58	Castor	Gemini (pg. 235)	52
19	1.62	Shaula	Scorpius (pg. 107)	700
20	1.64	Bellatrix	Orion (pg. 223)	240
21	1.68	El Nath	Taurus (pg. 231)	130
22	1.7	Alnilam	Orion (pg. 223)	1300
23	1.76	Alioth	Ursa Major (pg. 37)	81
24	1.79	Dubhe	Ursa Major (pg. 37)	120
25	1.8	Kaus Australis	Sagittarius (pg. 111)	140
26	1.82	Algenib	Perseus (pg. 177)	590
27	1.84	Wezen	Canis Major (pg. 227)	1800
28	1.85	Alkaid	Ursa Major (pg. 37)	100
29	1.86	Sargas	Scorpius (pg. 107)	270
30	1.9	Alhena	Gemini (pg. 235)	100
31	1.97	Polaris	Ursa Minor (pg. 41)	430
32	1.98	Murzim	Canis Major (pg. 227)	500
33	1.99	Alphard	Hydra (pg. 69)	180
34	2.0	Hamal	Aries (pg. 193)	66

GLOSSARY

Algol Variable: Star systems that include a large, bright primary star that is frequently eclipsed by a much dimmer companion star. Named after the star Algol in the constellation Perseus.

Andromeda Group: The group of constellations (Andromeda, Cassiopeia, Cepheus, Cetus, Perseus, and sometimes Pegasus) associated with the legend of Andromeda in Greek mythology.

Apparent Magnitude: A measure of the brightness of a celestial object as it is seen from Earth.

Asterism: A pattern identified in a group of stars belonging to one or more constellations. Examples include the Big Dipper, found in Ursa Major, and the Summer Triangle, formed by the brightest stars in Aquila, Cygnus and Lyra.

Asteroid: A body of rock in orbit around the sun with a mass too small for its gravity to compress it into a sphere.

Asteroid Belt: A region in space between the orbits of Mars and Jupiter where most of our solar system's asteroids orbit the sun.

Astrology: An ancient practice of using the positions of the stars and planets to predict the future, still used by some today.

Astrological Age: A time period in astrology that corresponds with one of the twelve signs of the zodiac. An astrological age is defined by the sign where the sun resides on the first day of spring. Each astrological age lasts for approximately 2,160 years.

Axis: A line through the center of an object, such as a planet or star, from its north pole to its south pole, around which the object rotates.

Bayer Classification: The system of naming stars with a Greek letter, followed by the name of the constellation. The brightest star in a constellation is listed as alpha, the second-brightest as beta, and so on.

Binary Star System: A star system that includes two stars that are gravitationally bound.

Black Hole: A celestial object that is infinitely dense with a gravitational field strong enough that not even light can escape it.

Circumpolar: A star, constellation, or other celestial object that never sets below the observer's horizon.

Comet: A celestial object with an icy, dusty nucleus (core) and which usually has an eccentric orbit around the sun. When it nears the sun, it heats up, vaporizing some of the ice. This vapor stretches out behind the comet, forming a "tail" that may extend tens of thousands of miles.

Constellation: A visual grouping of stars that defines a specific area of celestial sphere. Today, there are 88 official constellations recognized by the International Astronomical Union. Every point on the celestial sphere belongs to one, and only one, constellation.

Dwarf Planet: An object in orbit around a star that has enough gravity to compress itself into a sphere, but not enough to clear the path of its orbit of other objects. There are five currently recognized dwarf planets. This number will surely grow as astronomers discover more.

Ecliptic: The plane of the Earth's orbit projected onto the celestial sphere. It defines the apparent path the sun and the planets take across the sky throughout the year.

First Magnitude: A star with an apparent magnitude greater than 1.5, making it one of the brightest stars in the sky. The exact number of first-magnitude stars depends on how certain multiple star systems are counted.

Galactic Pole: The direction in our sky parallel to the axis of the Milky Way galaxy's rotation. When we look in these two directions, we are looking away from the disc of our own galaxy. The north galactic pole is located in Coma Berenices, while the south galactic pole is located in Sculptor.

Galaxy: Enormous collections of stars, gas and dust that are home to most of the visible matter in the Universe. Galaxies come in many shapes and sizes and can contain anywhere from a few million to over a trillion stars.

Giant Star: An unusually large, bright star. Generally, a star is considered to be a giant when it is 10–100 times the size and 10–1,000 times as bright as our sun.

Globular Cluster: A large, ancient cluster of stars that orbits a galaxy's core. Globular clusters often include hundreds of thousands of stars.

Great Square of Pegasus: A large asterism that dominates the fall sky that outlines the body of the horse in the constellation Pegasus. It includes a star from the constellation Andromeda, but is typically shown in its entirety in depictions of the constellation Pegasus.

International Astronomical Union (IAU): The internationally-recognized governing body for professional astronomy.

Light Year: The distance that light travels in one year, or about 5.88 trillion miles. A light year is a cosmic yardstick that helps put astronomical distances into perspective; the star closest to the sun is about 4 light years away.

Magnitude: The brightness of a celestial object. Brighter objects are assigned lower numbers and the brightest objects are given negative numbers. A first-magnitude star is roughly 2.5 times brighter than a second-magnitude star, which is 2.5 times brighter than a third-magnitude star, and so on.

Meteor: A small piece of rock, ice, dust or other debris falling to Earth from space and glowing white hot from friction with the Earth's atmosphere.

Meteorite: A meteor that does not completely burn up in the Earth's atmosphere and lands on the surface of the Earth.

Milky Way: Our home galaxy, the Milky Way is a relatively large spiral galaxy made up of about 200 billion stars. The light of billions of distant stars blend together to form an irregular band of soft light in our night sky.

Mira-class Variable Star: A class of variable stars; a Mira-class variable exhibits an apparent magnitude that varies at least 2.5 magnitudes every 100–1000 days.

Mythology: A culturally-specific collection of stories, legends and tales that revolve around gods, heroes and other legendary figures.

Nebula: A giant cloud of gas and dust in space. Diffuse nebulae are enormous clouds of interstellar gas and dust. Some are the birthplace of stars and a few are visible with the naked eye. Planetary nebulae are much smaller clouds of gas that form when a dying star throws off the outer layers of its atmosphere. They are called planetary nebulae because they look superficially like planets in a telescope.

Nuclear Fusion: The process that occurs when lighter elements such as hydrogen join together, or fuse, creating a heavier element and simultaneously producing a great deal of energy. Stars are powered by nuclear fusion.

Open Cluster: A group of stars that are loosely bound by gravity and move together through the galaxy. Generally, these stars share a common origin. Several open clusters are easily visible to the naked eye, including the Hyades and the Pleiades in the constellation Taurus.

Optical Double: A pair of stars that appear to be close together, when viewed from Earth. In reality these stars are quite far from one another and only appear to be close together.

Orion Nebula: A bright-diffuse nebula located in the constellation of Orion. An active star-forming region, it is clearly visible to the naked eye and a favorite object for amateur and professional astronomers alike.

Planet: An object in orbit around a star that has enough gravity to compress itself into a sphere and clear the path of its orbit of other objects. Most planets in our solar system have moons or satellites in orbit around them. Five of the seven other planets in our solar system can be readily observed with the naked eye.

Pleiades: A large, bright, open star cluster in the constellation Taurus. Recognized since ancient times, and once considered a separate constellation, it is one of three constellations named in the Bible. The cluster has been referred to as "seven sisters" for thousands of years.

Plutoid: A dwarf planet with an orbit that extends out beyond the orbit of Neptune.

Precession: The slow change or "wobble" of Earth's axis over long periods of time, precession is akin to the wobble of a top. Earth's "wobble" will eventually cause the axis to point in a different direction (thereby making a different star the "pole star").

Satellite: An object in orbit around another object. Usually refers to a natural or man-made object in orbit around a planet, dwarf planet or minor planet.

Star: A massive ball of super-hot gases, primarily hydrogen, that emits light and heat from nuclear fusion taking place in its core.

Summer Triangle: A large, prominent asterism in the summer sky that stretches across the band of the Milky Way. It is comprised of the first-magnitude stars Altair in Aquila, Deneb in Cygnus and Vega in Lyra.

Supergiant Star: An extremely large, bright star. Generally, a star is considered to be a supergiant when it is 30–1,000 times the size and 30,000–100,000 times as bright as our sun. Stars larger and brighter than supergiants are sometimes referred to as hypergiants.

Supernova: A cataclysmic explosion of a massive star that produces a wave of heat and light that sometimes outshines an entire galaxy and releases more energy than our sun will in its entire lifetime.

Variable Star: A star that exhibits fluctuations in its brightness over a time scale of hours to years.

Zodiac: An ancient series of twelve "signs" or "stations" that follow the ecliptic, tracing the apparent path of the sun relative to the constellations. Derives from the Latin *zōdiacus* meaning "circle of animals."

HELPFUL RESOURCES

Websites

Sky & Telescope (www.skyandtelescope.com)

The National Aeronautics and Space Administration (www.nasa.gov)

Sky Maps (www.skymaps.com)

The Astronomical League (www.astroleague.org)

Astronomy Magazine (www.astronomy.com)

The American Meteor Society (www.amsmeteors.org)

Clear Dark Sky (http://cleardarksky.com)

International Dark Sky Association (www.darksky.org)

The Constellations (http://www.dibonsmith.com)

Stellarium Planetarium Software (www.stellarium.org)

Books

Condos, Theony. *Star Myths of the Greeks and Romans: A Sourcebook*. Grand Rapids: Phanes Press, 1997.

Dickinson, Terence and Dryer, Alan. *The Backyard Astronomer's Guide, 3rd edition*. Richmond Hill: Firefly Books, 2008.

CONSTELLATION CHECKLIST

Use the boxes to check constellations you've seen.

INDEX

ABOUT THE AUTHOR

Jonathan Poppele is a naturalist, author and educator with wide-ranging interests. He earned a master's degree in Conservation Biology from the University of Minnesota, studying citizen science, environmental education and how to cultivate a personal relationship with the natural world. He has taught college level environmental studies, biology, and writing courses and currently teaches in the Department of Writing Studies at the University of Minnesota. An avid outdoorsman and student of natural history, Jon is a member of the Astronomical League, the Minnesota Astronomical Society, the Minnesota Trackers Club and the International Society for Professional Trackers. A Black Belt in the peaceful martial art of Ki-Aikido, Jon is the Founder and Director of the Center for Mind-Body Oneness in Saint Paul, MN. He can be reached through his school website at www.mindbodyoneness.com.